No. 3042
$22.95

The Mysterious Oceans

JON ERICKSON

*Discovering
Earth Science*

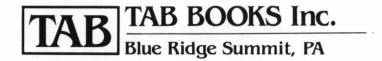

TAB BOOKS Inc.

Blue Ridge Summit, PA

FIRST EDITION
FIRST PRINTING

Library of Congress Cataloging in Publication Data

Erickson, Jon, 1948-
    The mysterious oceans / by Jon Erickson.

        p.   cm.
    Bibliography: p.
    Includes index.
    ISBN 0-8306-9142-1   ISBN 0-8306-9342-4 (pbk.)
    1. Ocean. I. Title.
GC21.E72 1988   88-11917
551.46—dc19   CIP

TAB BOOKS Inc. offers software for sale. For information and
a catalog, please contact TAB Software Department, Blue Ridge
Summit, PA 17294-0850.

Questions regarding the content of this book
should be addressed to:

    Reader Inquiry Branch
    TAB BOOKS Inc.
    Blue Ridge Summit, PA 17294-0214

Edited by Suzanne L. Cheatle
Series design by Jaclyn Saunders

Front cover: Great Barrier Reef photographed by Dr. Evon Barvinchack, amateur photographer and world traveler. Inset Earth photograph courtesy of NASA.

Contents

Acknowledgments

THE following organizations are recognized for their assistance in providing photographs for this book: the U.S Army Corps of Engineers, the U.S. Department of Agriculture (USDA), the National Aeronautics and Space Administration (NASA), the National Oceanic and Atmospheric Administration (NOAA), the U.S. Air Force, the U.S. Coast Guard (USCG), the U.S. Geologic Survey (USGS), the U.S. Maritime Administration (USMA), the U.S. Navy, and the Woods Hole Oceanographic Institution (WHOI).

Other Books in the
DISCOVERING EARTH SCIENCE SERIES

VOLCANOES AND EARTHQUAKES (No. 2842)

The first book in the series, this volume concentrates on the geologic phenomenon of Earth, and how this phenomenon has affected life on our planet. Also included is a history of the planet's geology.

VIOLENT STORMS (No. 2942)

This second book in the series deals with the atmospheric and climatic phenomena of Earth, as well as their effect on man and their influence on our planet.

THE LIVING EARTH (No. 3142)

The fourth book in the series focuses on the biologic phenomenon of Earth. It looks at the origins, history, and future of life, as well as the effects that the geologic, climatic, and hydrologic phenomena have on life.

EXPLORING EARTH FROM SPACE (No. 3242)

The final volume in the series, this book deals with the technologic advances that enable us to view our planet from space. It covers the way man-made satellites can provide us with better information on the geologic, climatic, biologic, and hydrologic phenomena of our planet and enable us to better predict disasters, locate and monitor natural resources, and explore our Solar System.

Introduction

FROM the viewpoint of earth scientists, our planet should be called *Ocean* because this is the only known planet that is literally drowning in water. Except for metallic mercury, water is the only mineral that exists naturally on the Earth's surface in the liquid form. About 70 percent of the surface of the Earth is covered by oceans, and 60 percent of the Earth's surface is covered by water over 1 mile deep. In the Pacific Basin, the ocean is 7 miles deep, and if Mount Everest, the tallest mountain, were placed there, the water would still extend over 1 mile above it. If the Earth were smooth and round like a billiard ball and completely covered by ocean, the water would have an average depth of almost 2 miles. Yet, in relation to the overall size of the Earth, the thin veneer of the ocean is practically insignificant and is comparable to the outer skin of an onion.

The ocean is criss-crossed by vast undersea mountain ranges greater than those found on land. Some of these mountains break through the surface of the water, and their peaks form islands. Active undersea volcanoes rise tens of thousands of feet from the ocean floor, creating new volcanic islands.

In this respect, the Hawaiian volcano Mauna Kea is the tallest mountain in the world. It rises 6 miles from the ocean floor, making it taller than Mount Everest by several hundred feet. Some canyons in the ocean floor are several miles deep and rival even the Grand Canyon.

There are strange life forms on the deep ocean floor that live in total darkness and near freezing temperatures. They congregate around hot-water vents from which they receive warmth and nutrients. They can survive only for as long as the vents continue to operate. There are mineral riches on the ocean floor just waiting to be plundered, and there is enough energy in seawater to power civilization well into the future.

The ocean floor is completely rejuvenated every 160 million years. New ocean floor is produced at midocean ridges, and old ocean floor is consumed at deep ocean trenches. Oceanic crust, seawater, and other substances are recycled through the mantle and reenter the Earth's atmosphere through volcanic vents. Volcanoes replenish the Earth with some of the vital ingredients it needs to stay alive. If the

Earth were no longer a tectonically active planet, it would be as dead as the Moon or Mars.

The ocean is a vast chemical factory, producing 80 percent of the Earth's oxygen, while scrubbing out carbon dioxide from the atmosphere and storing it in sediments on the ocean floor. While new ocean floor is being made by volcanic emissions at midocean ridges, the continents are being pushed wider apart in the Atlantic and closer together in the Pacific. In a couple of hundred million years, all the continents will come together in one single large land mass, and the world would be totally unrecognizable from what it is today.

Currents in the ocean distribute heat from the tropics to other parts of the world. The ocean's ability to store and move huge quantities of heat is far greater than the atmosphere's, and has a profound effect on the climate. The ocean retains the summer's heat and slowly gives it up in the winter, moderating the Earth's temperature between seasons. Furthermore, to change significantly the temperature of the upper 1000 feet of the ocean would take about a decade, and to change the temperature of the abyssal would take a thousand years or more.

The role of the ocean in shaping the weather has been recognized only recently. Every day, the Sun evaporates a trillion tons of seawater, which eventually returns to Earth as precipitation. Any elusive changes in the ocean and the atmosphere can lead to dramatic shifts in the behavior of the atmosphere, sending abnormal weather all around the world.

The ocean gave life to the Earth, and the first simple organisms evolved after only a billion years from the time the Earth formed out of the cosmic dust. A massive population explosion of photosynthetic organisms threatened to pollute the ocean with oxygen, a highly toxic substance to primitive life forms. Life was on the verge of extinction until respiration evolved, in which organisms took in oxygen for their growth.

Throughout the Earth's long history, there have been several episodes of mass extinctions of species because of some catastrophic environmental change. Every time an extinction occurred, life bounced back with more vigorous and complex species. After a tenuous layer of ozone formed in the upper atmosphere, which blocked out the Sun's deadly ultraviolet light, species were able to occupy the land. The plants took the first tentative steps out of the ocean and were followed by amphibious animals, who survived on the land and in the sea. The amphibians gave way to the reptiles and dinosaurs, who then gave way to the mammals.

No other species has the potential of making dramatic changes on the face of the Earth as does man. The attitude that the Earth is his for the taking has led to some profound and often destructive environmental impacts.

The release of pollutants into the atmosphere and the ocean is far greater and more widespread than ever before imagined. Chemical substances injected into the upper atmosphere are destroying the ozone layer in measurable amounts that could increase ultraviolet radiation, thus producing an increase of skin cancer. Overfishing has depleted the ocean of many species of good edible fish, leaving behind coarser trash fish, and certain species of whales are in danger of extinction. The need for more petroleum resources has led to oil spills that destroy nearby marine habitats. The ocean has become a dumping ground for enormous quantities of garbage, sewage, and hazardous wastes, all of which find their way into the marine food chain.

Industrialization also has played a part. Increased carbon dioxide brought on by the combustion of fossil fuels and the destruction of forests threatens to substantially raise global temperatures with the potential of melting the ice caps, raising the sea level, and flooding coastal regions. Acid rain is destroying the world's forests, acidifying freshwater lakes, and killing fish. As human populations continue to grow and to consume and poison the Earth on a grand scale, the damage could become irreparable, and the Earth might require millions of years to recover from man's folly.

1

Origin of Land and Sky

WHEN the Milky Way was about 10 billion years old, there was a mighty explosion in one of its spiral arms that was brighter than the entire Galaxy. In an instant, a giant star, a hundred times larger than the Sun, became a supernova and spewed its contents into the vast emptiness of space.

After traveling for some 10,000 years and 300 trillion miles, the gigantic shock wave encountered a diffuse cloud of gas and dust, called a *nebula* (FIG. 1-1). The shock wave completely enveloped the nebula and compressed it into a smaller volume. This compression caused the nebula to exceed a certain critical density, and it began to collapse upon itself under the influence of gravity. Debris from the supernova was injected into the collapsing nebula, causing it to rotate and flatten into a pancake-shaped disk. As the nebula continued to collapse, it rotated faster, creating spirals of matter that segregated into concentric rings. The center of the solar nebula collapsed into a dense glowing ball of hydrogen and helium. The heat generated by the gravitational compression started a thermonuclear reaction in the in-

terior, and after only 10 million years from the time the nebula first began to collapse, the Sun ignited.

BLUEPRINT FOR A SOLAR SYSTEM

About twice a century, there is a supernova somewhere in our Galaxy; therefore, supernovas might seem like rare events. If considered over the entire 15-billion-year life span of our Galaxy, however, they are actually quite numerous.

Over a period of several billion years, gas and dust from a multitude of supernovas accumulated in a diffuse protosolar cloud. Each supernova contributed its unique blend of elements, which were thoroughly mixed into the cloud. The protosolar cloud was composed of every natural element and its isotopes (chemically identical to the element, but with a different mass). Hydrogen and helium were the most abundant elements. Every hundred million years, density waves, which propagate around the center of the galaxy and are like the compressions and rarefactions of sound waves, passed in the neighborhood of the protosolar cloud. The density waves compressed portions of the cloud and triggered the formation of new stars. Some of these new stars were giants that only lived for a few million years before they became supernovas.

One of these supernovas happened to be within 60 light-years of a solar nebula that was in the early stages of star formation. The nebula was too diffuse for gravitational attraction to pull it together into a condensed state, but the close proximity of the supernova initiated the collapse of the nebula by the compressional forces from the huge shock wave. The nebula then exceeded the density necessary for self-gravitation, and it began to collapse. Debris from the supernova was injected into the collapsing nebula and imparted a counterclockwise rotation as viewed from Polaris, the North Star. This material was too late to mix thoroughly with that of the nebula, and instead, it condensed into a primitive class of meteors called *carbonaceous chondrites*, which are composed of carbon with small, round mineral in-

clusions and which are totally different from other chemical substances in our Solar System.

As the solar nebula continued to collapse, it rotated faster, just as an ice skater spins faster when she tucks in her outstretched arms. This faster rotation forced the nebula to flatten along a plane perpendicular to the axis of rotation. If the nebula had continued to collapse while it continued to gyrate wildly, it would have condensed into a star-cluster system of two or more stars, each revolving around a common center of gravity. This is the most common method of star formation. In the case of our Solar System, however, large centrifugal forces caused spiral arms to peel off, making the nebula look somewhat like a spiral galaxy.

The nebular material was then segregated into concentric rings, composed of primordial dust grains. In the center was a single compact, dense ball of hydrogen and helium, called a *protosun*. The heat generated by the gravitational compression made the

(Courtesy of NASA)

FIG. 1-1. The Orion Nebula.

protosun glow red hot. As additional material was attracted into the protosun by the strong gravitational force, it reached the critical temperature for thermonuclear reactions to take place in the core. In a flash, the Sun ignited and lit up the Solar System, which was in its first stages of development.

The transmutation of hydrogen into helium in the Sun's core generated copious amounts of thermal energy and a strong solar wind composed of atomic particles. The solar wind, along with gravitational torques, slowed down the Sun's rotation and transferred most of its *angular momentum*, or rotational energy, to its revolving disk of small, stony bodies called *planetesimals*. Thus, although it has 99.9 percent of the mass of the Solar System (FIG. 1-2), the Sun has less than 1 percent of the angular momentum.

A rapidly spinning sun could not have planets, at least not for very long. Assuming they formed at all, the planets would not be able to maintain their orbits and would spiral into the sun as a result of its large gravitational attraction. Fortunately, in our Solar System, nearly all of the angular momentum resides with the planets, mostly Jupiter. With the exception of Venus, the planets and most of their moons rotate on their axes in the same direction they revolve around the Sun, or counterclockwise. Although single stars are a minority in our galaxy, most of them can be expected to have planets because the formation of a planetary system is the natural consequence of the formation of a single star.

The inner terrestrial, or rocky planets— Mercury, Venus, Earth, and Mars—formed by the accretion of solid materials swept out of the solar nebula. The outer gaseous planets beyond the asteroid belt, which lies between Mars and Jupiter, formed by the gravitational attraction of large amounts of volatile substances, such as hydrogen, helium, water, ammonia, and methane. Well beyond the orbit of Pluto, the farthest known planet, is the Oort Cloud, from which originates the comets. Comets are hybrid planetary bodies composed of a stony inner core and an icy outer layer. They orbit the Sun at fantastic speeds, swinging close by the Sun then shooting back out into space in a highly eccentric orbit. Should the Earth get in the way of one

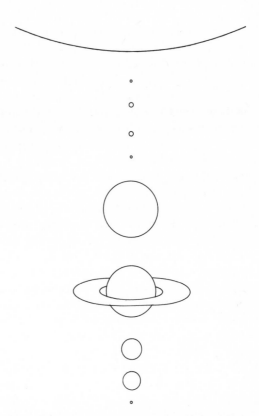

FIG. 1-2. The relative sizes of the Sun and planets.

or more of these icy visitors from outer space, the collision could prove to be catastrophic. They are the basis for one theory that accounts for periodic extinctions of species, such as the dinosaurs, every 26 million years.

MAKING A LIVING PLANET

The primordial dust grains in the solar disk stuck to one another by weak electrical attractions called *van der Waals forces*. These accumulations produced planetesimals of uniform dimensions about the size of a glass marble. The planetesimals swung around the Sun in elliptical orbits, which allowed constant collisions. The small bodies either rebounded off each other or adhered to each other by mutual gravitation attraction.

After 10,000 years, some bodies grew to over 50 miles in diameter, but most of the planetary mass still resided with the small planetesimals. If it were not for the presence of a large amount of gas in the solar nebula, the larger bodies would have continued to sweep up the remaining planetesimals, resulting in a solar system made up of thousands of planets, each smaller than the Moon. The Solar System would then resemble Saturn, with rings instead of planets. As it was, the gas slowed down the large bodies sufficiently for them to combine and form the nuclei of the terrestrial planets.

The first stages of the Earth's development was probably not much different from that of the other terrestrial planets. A constant bombardment of planetesimals, which converted their energy of impact into heat, kept the growing planets partially molten. In the presence of a gaseous medium, each planet attracted a massive initial atmosphere. The gravitational compression of the atmosphere of the growing Earth led to the melting of surface rocks long before the planet was completely formed. There also might have been some variation in the composition of materials in the solar nebula moving away from the Sun. This would have some bearing on the composition of the planets. In the Earth's case, there might have been abundant volatiles, including water, whereas Mercury, the closest planet to the Sun, was left high and dry. Also, Mars might have acquired an abundant amount of water as a result of greater distance from the Sun. Large quantities of water also could have been bound to hydrous minerals, which would release their water when the planet began to heat up.

After some 100 million years of formation, the Earth achieved nearly its present size. Its path around the Sun was swept clean of practically all planetesimals; the same is true for the rest of the terrestrial planets. The Earth was in a partially molten state from the inside out and glowed red hot as it journeyed around the Sun. On the surface, fountains of lava shot upward to great distances. Slabs of rock cooled, solidified, and dove into the hot interior to be remelted. Small rocky islands, or "rockbergs," floated on a sea of molten rock. Periodically, a large meteor plunged into the Earth, splashing up hot rocks and spreading them around for hundreds of miles. The gigantic crater soon filled up with magma, and no trace of the meteor remained behind.

Even in its earliest developmental stages, the young Earth was showing prospects of maturing into something special. Perhaps, it was its distance from the Sun of 93 million miles, give or take 1.5 million miles, or the tilt of its axis of 23.5 degrees, give or take 1.5 degrees, or the speed of its rotation which, incidentally, was much faster than it is today (FIG. 1-3). It might be that the composition of the Earth itself was responsible for making it unique among the planets.

GETTING TO THE CORE OF THE MATTER

There are two basic, but opposing, theories dealing with the origin of the Earth's core, as well as the rest of the planet. The oldest is the *homogenous model*, which holds that the accretion of nebular material, composed of both stony and metallic planetesimals, was undifferentiated. When the Earth grew to near its present size, short-lived and highly radioactive elements, called *radionuclides*, heated the interior as though it were one large nuclear core, similar to that of a nuclear generating station. The Earth's interior began a gradual meltdown because more heat was generated than was allowed to escape to the surface. There was also a large bombardment of meteors that heated and eventually melted the surface by impact friction. The Earth became one large molten sphere, and settling gravity caused the denser materials, particularly the abundant elements of iron and nickel, to migrate toward the center. Lighter elements floated toward the surface, forming what is called a basaltic *scum,* or primitive crust. When the short-lived radionuclides expended their energy, the Earth began to cool. Long-lived radionuclides—principally potassium, uranium, and thorium (called KUT for short)—took over and maintained the Earth's interior at a temperature that would allow convection currents in the core, and in the mantle above it.

According to the *nonhomogenous model,* the core formed first, and the mantle and surface rocks were gradually deposited on top of it. Getting small particles in the solar nebula to stick together becomes quite a problem because of the weak gravitational forces involved, and they might simply bounce

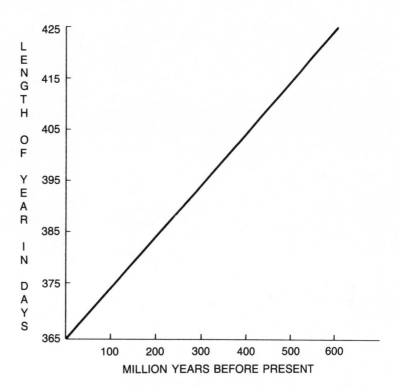

FIG. 1-3. Change in the Earth's rotation through time.

LENGTH OF YEAR IN DAYS

MILLION YEARS BEFORE PRESENT

off each other. If there were abundant ferromagnetic (iron and nickel) particles present, however, they would clump together by virtue of their mutual magnetic attraction. The magnetic particles could have come from a permanent magnetic core of one of the supernovas that provided the primordial material for the solar nebula. Even being magnetic is not all that necessary, and the metallic particles could have welded together by plastic deformation. By contrast, silicate materials would simply shatter upon impact. As the core grew near its present size, its large density created enough gravity to attract the rocky material out of the nebula. The more material that was added on top of the core, the greater became the force of gravity, which improved the Earth's ability to attract more material, until its path around the Sun was completely swept clean.

Both theories agree that the core heated up and differentiated into a liquid outer fraction and solid inner part. The inner core is composed of iron-nickel silicates and is roughly 1500 miles in diameter. The liquid outer core is mostly composed of iron with some carbon, silicon, and sulfur and has the viscosity

of molten iron. The core is 4320 miles in diameter, or a little over half the diameter of the Earth. The core makes up about one-sixth of the Earth's volume and about one-third of its mass. The density of the core varies from a minimum of about 9 times the density of water at the top of the core to about 12 times the density of water at the center. The temperature at the top of the outer core is about 4500 degrees Fahrenheit and increases to 5300 degrees Fahrenheit at the top of the inner core, with no appreciable increase toward the center. The pressure at the top of the outer core is 1.5 million *atmospheres* (atmospheric pressure at sea level) and increases to 3.5 million atmospheres at the top of the inner core, with only a slight increase in pressure toward the center.

Scientists are able to take pictures of the Earth's core through *seismic tomography*, a geologic version of the CAT scan. Like CAT scans, seismic tomography scans are produced by computers, but in this case, they combine information from earthquake waves that travel deep within the Earth's interior. By analyzing where the waves change speed

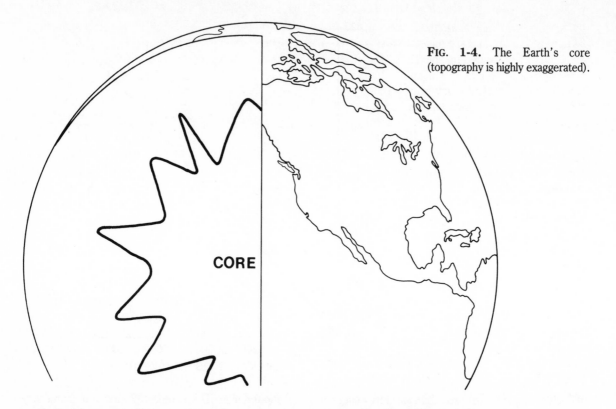

FIG. 1-4. The Earth's core (topography is highly exaggerated).

CORE

as they travel thousands of miles into the planet, scientists can determine where the mantle meets the core.

Seismic tomography has shown that the molten core is not round and smooth as it was once thought. Rather, it has mountains taller than Mount Everest and valleys deeper than the Grand Canyon (FIG. 1-4). There are peaks in the core 1800 miles beneath eastern Australia, the North Atlantic, Central America, south-central Asia, and the northeast Pacific, where a mountain rises 6 miles beneath the Gulf of Alaska. Valleys underlie Europe, Mexico, the southwest Pacific, and the East Indies, which lie above a canyon 6 miles deep. The peaks are created by rising hot currents in the mantle, which draw the core upward along with them. The valleys are formed when cold material in the mantle sinks and presses against the core, creating abysses. The mountains and valleys probably last only as long as it takes one current to rise and fall, or roughly 100 million years.

IN THE MIDST OF THINGS

Scientists know more about the far reaches of outer space than they do about the interior of the Earth. Even the deepest drill holes only penetrate at most a few miles into the Earth's thin membrane, and none has ever reached the mantle. Project Mohole, sponsored by the National Science Foundation, made a daring attempt to drill through the ocean crust in the mid-1960s, but Congress cut off funding millions of dollars and thousands of feet short of the goal.

Volcanoes erupt molten rock that originated deep within the earth, but the source material is not the mantle itself, but recycled ocean crust. Even if mantle material were directly involved, as it might be with hot-spot volcanoes, its composition would change by the assimilation of crustal rocks along its way to the surface. Basalts that erupt on the ocean floor at midocean ridges might bear close chemical resemblance to the upper mantle, but they do not offer any clues about the crystalline structure of the mantle rocks. Kimberlite pipes, which are another form of volcanism and are mined extensively for diamonds, have coughed up special rocks called *kimberlites* from as deep as 150 miles below the surface. Even at this great depth, however, they only scratch the surface of the mantle, which has a total thickness of 1800 miles.

The composition of the mantle can be surmised by analyzing the chemical content of stony meteorites. Also, moon rocks brought back by the Apollo lunar missions of the early 1970s (FIG. 1-5) have given scientists an insight about the chemical makeup of the mantle.

The majority of the Earth is mantle, which accounts for nearly 50 percent of the radius, 83 percent of the volume, and 67 percent of the mass of the planet. The mantle starts at an average depth of about 25 miles below the surface. The density of the mantle increases with depth from 3.5 times the density of water near its surface to about 5.5 times the density of water near the core. The temperature ranges from 2000 degrees Fahrenheit, at the top of the mantle to 4500 degrees Fahrenheit near the bottom. The mantle is composed of iron and magnesium-rich silicate minerals, corresponding to olivine in the uppermost part and to pyroxene and garnet, which make up the bulk. The rocks are in a partially molten or plastic state, which allows them to flow but only very slowly.

The mantle played a very important role in shaping the Earth and giving it a unique characteristic. Early on, large amounts of gases and water vapor sweated out of the mantle through volcanic pores in what is called the ''big burp.'' The mantle provided the oceans, the atmosphere, the rocks to build the continents, and the carbon compounds from which life was created.

The mantle is a great heat engine, in which the solid rock of the Earth's interior churns over ever so slowly, carrying the fiery heat of the inner Earth toward the surface. The mantle is also stratified into various layers, ranging from the lower mantle, which is depleted in the radioactive and crust-forming elements, to the upper mantle, which is rich in residual fluids. Convection in the mantle extends in one continuous turbulent cauldron from the top of the core to the bottom of the crust. These large-scale convections in turn drive small-scale convections, or

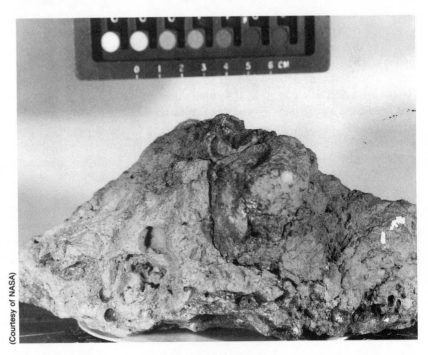

FIG. 1-5. A Moon rock.

(Courtesy of NASA)

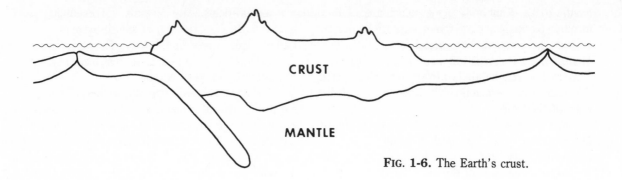

CRUST

MANTLE

FIG. 1-6. The Earth's crust.

cells, in the upper mantle, which are responsible for moving the continents around the surface of the globe. This activity created most of the land features on the Earth, from tall mountain ranges to deep ocean trenches. It is this active convection that keeps the planet alive; for without it, the Earth would become as tectonically and biologically dead as the Moon and Mars.

THE CRUST MAKES THE DIFFERENCE

The crust is often depicted as leftover material the mantle did not want, or scum that floated to the surface like the slag on the surface of molten iron ore. The Earth began to cool down soon after it differentiated into its various layers by gravity settling. The Earth no longer had its massive atmosphere to keep in the heat, and its interior quickly lost heat to space. At first, just a thin film formed on the surface, like the film on top of a cooling bowl of pudding. This film checked the escape of heat except for areas where the thin crust was temporarily remelted from below. As the crust continued to cool and thicken, it shriveled up like the skin of a baked apple. It then broke apart into several crustal plates that floated around the surface, crashing into each other much like pack ice does in the polar oceans. There is no vestige of the original crust, which might have gotten so thick it remelted, or became so unstable it overturned like a top-heavy boat. This might be the reason why there is no record of the first 700 million years of the Earth's geologic history.

The present crust is composed of the light elements of the Earth, with oxygen, silica, and alumi-num comprising over 80 percent of the total. These basic elements form granitic rocks that make up the bulk of the continents. The continents exist only because they are more buoyant than the mantle. Like icebergs, only the tip of the continents shows through, while most of the continents remain buried deep in the fluid, upper mantle (FIG. 1-6). The continental crust is 25 to 30 miles thick in most places, and it is the thickest (45 miles) along the Himalayan Mountain chain.

The ocean crust is much thinner. In places on the ocean floor, the crust is only 3 to 5 miles thick. The ocean crust is composed of a heavier iron and magnesium-rich rock called basalt. The continental crust is 20 times older than the ocean crust, and in no area is the ocean crust older than 200 million years.

The lithosphere, upon which the continents and the ocean floor ride, is not a solid shell like the shell of an egg, but is broken into a dozen major plates and several minor ones. These plates move about freely on a sea of molten rock called the *asthenosphere*. The plates interact with each other either by extension, collision, or lateral stress.

The plates move away from each other at midocean ridges, where new ocean floor is constantly being produced. This is why ocean crust is so much younger than continental crust. Where two plates collide, the heaviest is subducted under the lightest, creating a deep ocean trench. In a collision between an oceanic plate and a continental plate, the oceanic plate is always subducted under the continental plate. When two oceanic plates collide, one is subducted under the other. When two plates slide

TABLE 1-1. Classification of the Earth's Crust.

ENVIRONMENT	CRUST TYPE	TECTONIC CHARACTER	CRUSTAL THICKNESS	GEOLOGIC FEATURES
Continental crust overlying stable mantle	Shield	Very stable	22 miles	Little or no sediment; exposed Precambrian rocks
	Midcontinent	Stable	24 miles	Little or no sediment; exposed Precambrian rocks
	Basin-range	Very unstable	20 miles	Recent normal faulting, volcanism, and intrusion; high mean elevation
Continental crust overlying unstable mantle	Alpine	Very unstable	34 miles	Rapid recent uplift, relatively recent intrusion; high mean elevation
	Island arc	Very unstable	20 miles	High volcanism; intense folding and faulting
Oceanic crust overlying stable mantle	Ocean basin	Very stable	7 miles	Very thin sediments overlying basalts; no thick Palaeozoic sediments
Oceanic crust overlying unstable mantle	Ocean ridge	Unstable	6 miles	Active basaltic volcanism; little or no sediment

TABLE 1-2. Composition of the Earth's Crust.

CRUST TYPE	SHELL	AVERAGE THICKNESS IN MILES	PERCENT COMPOSITION OF OXIDES						
			SILICA	ALUM	IRON	MAGN	CALC	SODI	POTAS
Continental	Sedimentary	2.1	50	13	6	3	12	2	2
	Granitic	12.5	64	15	5	2	4	3	3
	Basaltic	12.5	58	16	8	4	6	3	3
Total		27.1							
Subcontinental	Sedimentary	1.8							
	Granitic	5.6			Same as above				
	Basaltic	7.3							
Total		14.7							
Oceanic	Sedimentary	0.3	41	11	6	3	17	1	2
	Volcanic sedimentary	0.7	46	14	7	5	14	2	1
	Basaltic	3.5	50	17	8	7	12	3	<1
Total		4.5							
Average		15.4	52	14	7	4	11	2	2

past each other, they form transform or lateral faults. The San Andreas fault in Southern California (FIG. 1-7) is a good example of a lateral fault where the Pacific plate is sliding northward past the North American plate.

The Earth's crust is the most unique among the terrestrial planets. The Moon, with only a quarter of the Earth's diameter, has a crust of roughly the same thickness. Although the Moon and Mars were volcanically active in their early history, they now appear to be tectonically dead. This could be the result of rapid cooling and development of a lithosphere that was too thick and too buoyant to break apart and subduct into the mantle. Venus, which is about the same size as the Earth, was thought to be its geological twin. Because of the high surface temperatures resulting from an atmosphere almost entirely composed of carbon dioxide, the lithosphere on Venus should be thinner, weaker, and more mobile than that on Earth. However, Venus lacks any signs of this activity such as linear ridges or trenches. Apparently, both low pressures found on Mars and the Moon and high temperatures found on Venus favor the development of a thick buoyant crust that cannot be subducted.

It is often suggested that life originated on Earth because the extremes in surface temperature are within the narrow temperature range of liquid water. There is also an interesting possibility that the crust remains mobile because the Earth has life. Limestone-secreting organisms in the ocean scrub out carbon dioxide from the atmosphere, which keeps the Earth from having a runaway greenhouse effect like that on Venus, whose crust is unable to renew itself.

THE EARTH'S SISTER PLANET

Early in its history, when the Earth was still in a molten state and without an atmosphere, it was constantly being bombarded by meteors, some possibly as large as 50 miles in diameter. The larger meteors impacted with enough energy to send rocks flying 100 or more miles into space, but none of them ever reached escape velocity and returned to Earth. Then out of the heavens came a wayward asteroid about the size of Mars, traveling about 20,000 miles per hour. It struck the Earth with a glancing blow. A direct hit at this velocity would probably have shattered the Earth into thousands of meteor-sized

FIG. 1-7. The San Andreas Fault, California.

(Photo by R.E. Wallace, courtesy of USGS)

bodies—a fate that might have happened to a planet between Mars and Jupiter. The impact resulted in a gigantic explosion, and mantle material from the Earth and material from the asteroid were partially vaporized. As the asteroid sped away back into space, its gravitational attraction pulled jets of this material along with it. Some of the material was sent out into space, and some of it returned to Earth. A good portion, however, was sent into orbit around the Earth, forming a prelunar disk. A protomoon then began to form by the accretion of matter from the prelunar disk.

This scenario for the creation of the Moon tidies things up a little by taking the best of the other theories, including rotational fission, capture, and binary accretion, and making them part of one cohesive hypothesis. Creating the Moon out of the Earth's mantle material might account for why, of all the terrestrial planets, the Moon has little or no iron in its core and consists almost entirely of rock. The difficulty lies in proving that Moonrocks brought back by the Apollo missions are the same as the earth's mantle rocks since there is no means at the present time to sample the mantle directly. Also, because the Moon has been tectonically dead for 2 billion years, while the Earth remained tectonically alive, the mantle rocks would have changed dramat-

ically over the intervening period.

When the Moon was fully formed, it was about half the distance it presently is from the Earth. The Earth rotated much faster on its axis and therefore retained a large amount of the angular momentum of the Earth-Moon system. Early on, the Moon caused much larger tides to rise in the crust, the ocean, and the atmosphere than it does today. Tidal currents, dragging across the bottom of shallow seas, dissipated tidal energy as frictional heat, which slowed down the rotation of the Earth. This imparted more angular momentum to the Moon and sent it out into a wider orbit.

Presently, the Moon is receding from the Earth at a rate of about 2 inches per year. The Moon does not orbit the Earth in a perfect circle, but has an elliptical orbit with a perigee of about 221,000 miles and an apogee of about 253,000 miles. The orbital time from perigee to perigee is about 27.5 days. During a new moon or a full moon, each with a synodic period of 29.5 days, the Moon and Sun pull together on the Earth, causing the highest tides, which can lift the ocean several tens of feet. It is a distinct possibility that by having a nearby Moon to pull on the oceans, causing a lot of sloshing of water, the Earth acquired one of the most important conditions for the creation of life.

2

The Super Ocean

FROM the very beginning, there was every indication that something was special about the Earth (FIG. 2-1). Throughout its formative years, the Earth had more going for it than the other planets did. Maybe it was its particular position in the Solar System. Then again, it might have been its size, since it is the largest of the terrestrial planets. Perhaps its density was a critical factor since gravity which is dependent on a planet's mass is important to its development.

The Earth had much of its mass in a heavy iron-nickel core, which created a strong magnetic field that deflected solar particles. It had a mobile mantle composed of semimolten rock and volatiles, mostly water. The water escaped from the mantle through constantly erupting volcanoes. Some also rained down from outer space by a massive shower of icy meteorites that peppered the young planet during its early development. The Earth, by virtue of its distance from the Sun, developed and was able to maintain an atmosphere rich in oxygen and an abundant supply of water. These two factors are what essentially set the Earth apart from the rest of the planets.

PUTTING ON AIRS

For the first 0.5 billion years while it was spinning rapidly and surface rocks were still scorching hot, the Earth was without an atmosphere and the surface was in a near vacuum, much as the Moon is now. Without an atmosphere or oceans to distribute the Sun's heat, the Earth's surface baked at the temperature of molten iron during the day and froze to more than 100 degrees below zero at night. During this time, both day and night were only a few hours long, and the Moon loomed so close that it filled much of the sky. Soon after the start of the meteor barrage and as a result of it, the Earth slowly began to acquire a new atmosphere. Some of the meteorites were composed of rock and metals; some were composed of frozen gases and water ice; and others contained carbon as though millions of tons of coal were raining down from the heavens. Comets, which are essentially rocks encased in ice, also plunged into the Earth, releasing large amounts of water vapor and gas. These cosmic gases were mostly carbon dioxide, ammonia, and methane.

Most of the water vapor and gases came from within the Earth itself. *Magma*, which is the molten rock supplied to volcanoes, contains large amounts of volatiles, mostly water, which make it more fluid. The volatiles remain in the magma while it is deep inside the earth because tremendous pressures keep the volatiles from escaping. When the magma rises to the surface, however, the drop in pressure releases the trapped water and gases explosively. The early volcanoes erupted thousands of times more violently than the largest volcanoes of recorded history because the Earth's interior was much hotter, and the magma contained more volatiles than it does today. Modern volcanoes spew out large quantities of water vapor, carbon dioxide, nitrogen, and sulfur gases (FIG. 2-2), and it is not unexpected for them to have erupted in the same manner in the past.

Oxygen was produced either directly by outgassing from volcanoes and degassing by meteorites or indirectly by the disassociation of water vapor and carbon dioxide by ultraviolet light from the sun. Any oxygen that formed was quickly bonded to metals in the crust in the same manner that oxygen reacts with iron to make rust. Oxygen also recombined with hydrogen and carbon monoxide to reconstitute water vapor and carbon dioxide.

It is fortunate that the early Earth had very little molecular oxygen in its atmosphere; if it had, life could not have been created because oxygen would have been poisonous to primitive life forms. Nitrogen, which is practically inert, came from volcanic eruptions and the disassociation of ammonia, a molecule of one nitrogen atom and three hydrogen atoms. Unlike the other gases, which have been

(Courtesy of NASA)

FIG. 2-1. View of the Earth from Apollo 17.

recycled many times through the ocean, the Earth still contains much of its original nitrogen. The Earth's early atmosphere thus contained nitrogen, carbon dioxide, ammonia, methane, and water vapor. It was so saturated with water vapor that the atmospheric pressure was several times greater than it is today, making the atmosphere totally different from the one we have now.

WHENCE CAME THE RAIN

Ancient metamorphosed, water-lain sediments have been found and dated at 3.8 billion years old; therefore, the Earth must have had surface water sometime before the formation of these sediments. During the intervening 200 million years between the last of the great meteor bombardments and the first formation of sedimentary rocks, the Earth's surface was flooded with vast quantities of water. At first, the crust was still too hot to allow standing bodies of water, and any rain that fell simply evaporated before it reached the ground. The surface temperature was maintained at a high level, mostly as a result of the greenhouse effect. Sunlight was still feeble compared with what it is today, but the greenhouse gases—mostly carbon dioxide, methane, and water vapor—held in what little heat the Earth received from the Sun. Opposing this greenhouse effect were thick clouds of condensed water vapor that completely shrouded the planet. The clouds reflected much of the Sun's rays back into space by what is called the albedo effect (FIG. 2-3). The clouds also kept the heat that radiated from the hot

(Courtesy of USGS)

FIG. 2-2. The May 18, 1980, eruption of Mount St. Helens.

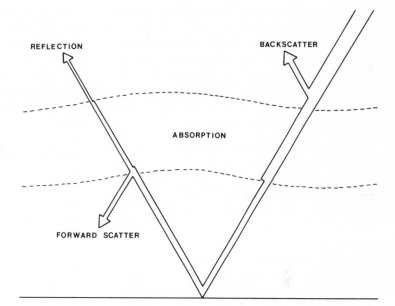

REFLECTION

BACKSCATTER

ABSORPTION

FORWARD SCATTER

FIG. 2-3. The effect of the albedo on incoming solar radiation.

crust from escaping into space by absorbing it and reradiating it back to the surface.

The entire surface of the Earth was in a rage. Winds blew with such a force they make present-day hurricanes seem like gentle breezes. Dust storms were prevalent on the dry surface and covered the entire planet with suspended dust, much like the Martian dust storms do today. Huge lightning bolts, carrying many times the electrical charge of those of today, flashed constantly from cloud to cloud and from cloud to ground. Upon striking the ground, the powerful lightning exploded like a large bomb, gouging out tons of rock. The thunder was earth-shattering as one gigantic shock wave after another compressed and rarified the air.

In addition to these awesome meteorological displays, volcanoes erupted in one giant outburst after another. The sky lit up from the pyrotechnics created by the white-hot sparks of ash and the eerie glow from the flow of red-hot lava. The restless Earth was rent apart by massive earthquakes that cracked open the thin crust. Magma flowed through the fissures and paved over the surface with lava, forming flat featureless plains, like the maria on the Moon and Mars.

Intense volcanic activity lofted millions of tons of volcanic ash and dust particles into the at-

mosphere. When there was no atmosphere to impede its fall, ejecta from volcanoes simply fell back directly around the volcanic vents and built cones up to prodigious heights, several times taller than the tallest volcanoes today. After the atmosphere was developed, the volcanic ash and dust remained suspended for long periods.

The dust acted as nuclei upon which water vapor could condense. When the upper atmospheric temperature lowered to the dew point, water vapor condensed into cloud droplets that were thousands of times smaller than a normal raindrop. Clouds became so thick and heavy they blocked out the Sun, and the Earth was in near darkness. The lack of sunlight allowed the surface to cool below the boiling point of water. Further cooling of the atmosphere allowed raindrops to form, and the Earth received deluge after deluge. Raging floods, the size of which make today's floods seem like small rivulets, cascaded down steep mountain slopes and gouged out deep canyons in the rocky plain. The rain first collected in meteor craters and volcanic calderas, which then overflowed and combined into small shallow seas.

When the rains abated and the skies finally cleared, the Earth was transformed into a giant blue orb, covered almost entirely by an ocean up to 2 miles deep. Any volcanoes that did manage to break the surface of the sea formed small volcanic islands. These islands quickly disappeared when their peaks eroded below sea level by fierce wave action, and they became flat-top, undersea mountains called *guyots*.

THE UNIVERSAL SEA

The bottom of the new ocean was in stark contrast with what it is today. Barren of life and without sediments deposited on the seafloor, the geography of the ocean bottom consisted almost entirely of scattered, jagged undersea volcanic mountains and monotonous lava plains. New volcanoes sprang up from the ocean floor, some rising tall enough to touch the sky, but were just as quickly eroded below sea level. Undersea earthquakes cracked open the thin crust, which was only two or three miles thick in most places. Lava bleeding from these underwater fissures piled up layer upon layer of basaltic rock.

The volcanoes and fissures were also escape valves for steam and gases. In these volatiles were compounds of carbon, sulfur, nitrogen, phosphorous, hydrogen chloride, sodium, and calcium, just to name a few. Seawater also seeped into cracks in the crust, where it was heated, leached water-soluble compounds out of the rock, and rose to the ocean floor again as undersea geysers. These ingredients quickly turned the sea into a vast chemical soup from which life would eventually emerge.

The climate, which was much warmer than it is today, supported the evaporation of large quantities of water from the ocean. The water vapor formed huge billowing clouds that raced across the planet, pulled along by the swift winds. Huge circulating weather systems were generated by a strong Coriolis effect. This force, as a result of the Earth's rotation, deflects air currents to the right in the Northern Hemisphere, and to the left in the Southern Hemisphere.

Ocean currents, unimpeded by land, evenly distributed the Sun's energy around the world. The Coriolis effect, along with greenhouse gases in the atmosphere, maintained the Earth at nearly the same temperature from pole to pole. The ocean also was heated from the bottom up by active volcanism on the ocean floor. The volcanism thickened the oceanic crust and continually supplied the seawater with life-giving nutrients.

The surface of the ocean was in constant agitation. Huge, choppy waves were stirred up by strong winds that blew steadily, dragging on the water and pulling it along with them. In the tropics, several gigantic hurricanes, much more powerful than the most destructive hurricanes today (FIG. 2-4), were constantly on a rampage across the planet because there was no land to impede their travel. The shifting of the crust produced huge earthquakes on the ocean floor, which sent great seismic sea waves around the world. The nearby Moon, with its large gravitational pull, caused the ocean to bulge in its direction as it rotated around the Earth. When the Sun and the Moon both pulled together on the Earth, the ocean responded with a violent upheaval of water that had an effect at great depths. All this agitation stirred the ocean as though it were a vast cooking pot, and all its ingredients were thoroughly mixed.

THE SUPERCONTINENT

The continents today are composed of odds and ends of ancient continental crust called *cratons*. Within these cratons are the oldest rocks found on Earth, dating back to 3.8 billion years. These rocks are composed of highly altered granite, as well as metamorphosed marine sediments and lava flows. The rocks originated from intrusions of magma into the primitive ocean crust. The magma slowly cooled and separated into a light component and a heavy component, which settled to the bottom of the magma chamber. Some of the magma also seeped through the crust, where it poured out as lava on the ocean floor. Successive intrusions and extrusions built up new crust until it finally broke the surface of the water. Unlike the volcanic islands, which existed for only a short while, these thin slivers of land became a permanent part of the landscape because they were lighter and more buoyant than the ocean crust. If a craton was forced into the mantle under the pull of gravity, it would bob up again like a cork on a fishing line.

The cratons, which numbered in the dozens, slowly built up and ranged from about one-fifth the size of present-day North America to a lump smaller than the state of Texas. The total land area was probably no more than one-tenth what it is now. The cratons also were very mobile and moved about freely on the molten rocks of the asthenosphere. They became independent, free-wheeling minicontinents that periodically collided with and rebounded off each other. The collisions caused a slight crumpling at the leading edges of the cratons, forming small parallel mountain ranges perhaps only a few hundred feet high. Volcanoes were highly active on the cratons, and lava and ash continued to build them upward and outward. New crustal material also was added to the cratons from magmatic intrusions composed of molten crustal rocks that were recycled through the upper mantle. This process effectively cooled the mantle, and the cratons slowed their erratic wanderings.

As the cratons became more sluggish, they developed a greater tendency to stick together when they collided. The point at which the cratons adhered to one another forced up mountain ranges, and the sutures joining the landmasses are still visible today as cores of ancient mountains over 2 billion years old. All the cratons eventually coalesced into a single large landmass several thousand miles wide. This was a strange-looking world of rock, looking much like the surface of Mars and totally devoid of life; for as yet, there were no land plants. They would not put in their appearance for another 1.5 billion years.

Weathering and sedimentary processes were much more active because there was no protective cover, which plants provide today. Rivers choked on detritus and changed their courses often as they flowed toward the sea. When the rivers reached the sea, they dropped their load of sediments, which continually built up the continental margins.

(Courtesy of NASA)

FIG. 2-4. Typhoon Bill from the space shuttle.

The interior of the supercontinent cracked open in many places, and the continent spread apart like the East African rift system is doing today, allowing the sea to rush in and invade the land. The inland seas eventually became choked with sediments, and as new land was uplifted, the seas were forced out. Sometimes the rifting formed large gulfs, and when the land came together again, mountains were pushed up. This constant rifting and patching up of the interior and depositing of sediments along the continental margins eventually built the continent outward until its area was nearly equal to the total area of all the present-day continents.

Regardless of its immense size, the supercontinent still roamed freely across the face of the Earth. Two billion years ago, it hovered over one of the polar regions, and the Earth experienced its first great ice age as ice over 1 mile thick covered much of the continent. This was also a period of transition when the atmosphere, which contained a large amount of carbon dioxide, was being replaced by one containing oxygen generated by photosynthesis in the ocean. The loss of carbon dioxide caused the climate to cool and encouraged the growth of ice.

Around 700 million years ago, there existed the greatest ice age the Earth has ever known. Around 600 million years ago, the supercontinent broke in two, marking the beginning of another transition period called the Phanerozoic eon. During this time, there was an explosion of animal life in the ocean.

The southern landmass resulting from the division of the supercontinent was called *Gondwanaland*, named after the Gondwana Province of India. It consisted of present-day Africa, South America, Australia, Antarctica, and India. India later broke off of Gondwanaland and became a subcontinent of Asia. The northern landmass was called *Laurasia*, named for the Laurentian Province of Canada and the continent of Asia. It consisted of the present landmass of the Northern Hemisphere. The sea separating Gondwanaland and Laurasia was called the Tethys Seaway, and Europe and North America were temporarily separated by a proto-Atlantic Ocean. Gondwanaland moved into the south polar region, where ice sheets covered large portions of the landmass.

About 240 million years ago, Gondwanaland and Laurasia rejoined in a great continental collision, and successive bumps and grinds created many of the world's mountain ranges. This new supercontinent was called *Pangaea*, meaning all lands. It was crescent-shaped and stretched practically from pole to pole. Pangaea was surrounded by a superocean called *Panthalassa*, meaning universal sea.

BREAKING UP THE MARRIAGE

About the time the dinosaurs began to roam the Earth around 180 million years ago, a great rift ran down the middle of Pangaea between North America and Europe and between South America and Africa. The rift soon filled with sea, which became the young Atlantic Ocean (FIG. 2-5). At first, the continents moved apart much faster than they do today, possibly as much as one-half foot per year.

Land bridges, or isthmuses, allowed the migration of animals between the continents until they became too far apart. As the continents of North and South America raced across the ocean on their way to their present positions, their leading edges crumpled into the long, sinuous mountain ranges of the Rockies in North America and the Andes in South America.

By the time the dinosaurs became extinct 65 million years ago, three major bodies of water were formed: the Atlantic, the Arctic, and the Indian oceans. The seas also invaded the interior of most continents, forming large inland seas (FIG. 2-6). The inland seas filled with sediments, and subsequent uplifting drove out the waters, leaving only salt lakes behind.

During the breakup of Pangaea, volcanic activity was intense and widespread. The higher content of greenhouse gases in the atmosphere caused the climate to warm considerably. Most of the continental landmass was located in the tropics where it could absorb more sunlight. There is no evidence of ice ages during this time because the polar regions were free of land, which tends to prevent warm ocean currents from keeping the poles ice free. The warm climate might have been one of the factors that led to the giantism of the dinosaurs. Today, huge coal deposits testify that during this time, know as the

TABLE 2-1. The Drifting of the Continents.

PERIOD OR EPOCH	AGE*	GONDWANALAND	LAURASIA
Quaternary	3		Opening of Gulf of California
Pliocene	11	Beginning of spreading near Galapagos Islands	Change of spreading directions in eastern Pacific
		Opening of the Gulf of Aden	Birth of Iceland
Miocene	25	Opening of Red Sea	
Oligocene	40	Collision of India with Eurasia	Beginning of spreading in Arctic Basin
Eocene	60	Separation of Australia from Antarctica	Separation of Greenland from Norway
Paleocene	65	Separation of New Zealand from Antarctica	Opening of the Labrador Sea
			Opening of the Bay of Biscay
		Separation of Africa from Madagascar and South America	Major rifting of North America from Eurasia
Cretaceous	135	Separation of Africa from India, Australia New Zealand and Antarctica	
Jurassic	180		Beginning of separation of North America from Africa
Triassic	230		
Permian	280		

*in millions of years

Cretaceous period, vegetative growth was prodigious and provided a substantial diet for the giant reptiles.

The warmer climate also drove more intense weather features, and some mountain ranges were quickly eroded to nubs. Meanwhile, new mountains were formed by continents pressing against each other. Gibraltar uplifted and formed a dam that blocked the waters of the Atlantic from entering the Mediterranean Sea. The sea completely evaporated,

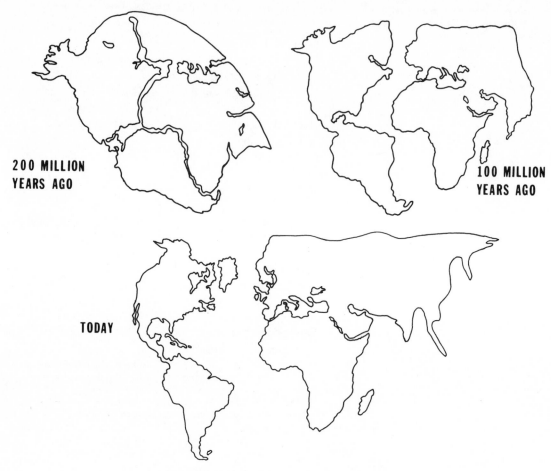

FIG. 2-5. The opening of the Atlantic Ocean.

creating a huge, dry basin. Eventually, the dam subsided, the Atlantic poured into the basin, and the Mediterranean became a sea once again.

The movement of bodies of land across the globe also affects the shapes of the ocean basins. The ocean bottom influences how much heat is carried by the ocean currents from the tropics poleward. When Antarctica separated from South America and later Australia about 40 million years ago, and moved into the south polar region, a circumpolar antarctic ocean current was formed. This current isolated the frozen continent and prevented it from ever obtaining warmth from poleward-flowing waters, thereby condemning it to a life of ice until it moves out of the polar region once again. Since

the extinction of the dinosaurs at the end of the Cretaceous period, the ocean bottom has grown steadily colder, so that now it is near the freezing point of water (FIG. 2-7).

The Arctic Ocean is the only ocean in the world that is practically landlocked. The land effectively blocks warm currents of the tropics from reaching the North Pole and melting the polar ice cap.

The Pacific Ocean will continue to shrink as North and South America march across the Pacific Basin. The isthmus connecting the Americas will drown as the two continents continue to pull apart, and ships will no longer need the Panama Canal. Southern California will slice its way northward, where it will disappear down the Aleutian trench.

FIG. 2-6. The Middle Jurassic inland sea (top). The Upper Cretaceous inland sea (bottom).

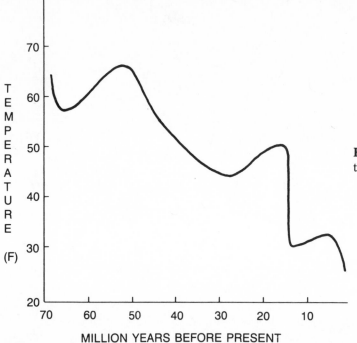

FIG. 2-7. Ocean-bottom temperature through time.

Africa and Eurasia will continue to press against each other, and the Mediterranean Sea, caught in the middle, will be squeezed dry. A new subcontinent will break free from East Africa and collide with India. Australia will continue to drift northward and crash into Southeast Asia. As time goes on, all the continents will come together to form a new supercontinent, called *Neopangaea*. The supercontinent will eventually break up as it has done before. The new continents will bear no resemblance to what they are today, and the Earth will become a strange new world.

3

The Great
Chemical Factory

AFTER the surface cooled and the ocean formed, life on Earth could hardly wait to get started. Conditions on Earth encouraged the formation of the precursors of life, and the planet acquired life much earlier than it ever though possible (FIG. 3-1). In the mid-1950s, the American chemists Stanley Miller and Harold Urey conducted an experiment with a special apparatus that was designed to represent conditions on the early Earth. A flask of boiling water represented the primordial sea, and another container holding hydrogen, methane, ammonia, and water vapor represented the early atmosphere. The gases were passed over an electric spark to simulate lightning or the strong ultraviolet light that easily penetrated the early oxygen-free atmosphere. The gases were then cooled, allowed to condense, and collected in a U-shaped trap. Meanwhile, water was continuously boiled in the flask and recirculated through the spark chamber and condenser in closed-loop system. After the apparatus was allowed to cook for about a week, a dark soup was collected from the trap and was found to contain complex amino acids—the very building blocks of life.

LIFE AMONG THE MUD FLATS

When the rains first came and began to wash over the hot crust some 4 billion years ago, they dissolved substances that were created by chemical reactions in the atmosphere and in the crust and deposited them in primitive shallow seas. The seawater evaporated, formed clouds, and precipitated as rain, mostly into the sea again. A continuous circulation between the ocean, the atmosphere, the land, and back to the ocean, along with the circulation of water through the ocean crust, provided the seas with an abundance of chemical substances (FIG. 3-2). Chlorine ions from outgassed hydrogen chloride combined with sodium ions leached from rock to make sodium chloride, or common table salt, and over time, the seas acquired considerable saltiness.

Earlier theories of evolution relied on the "primordial soup" hypotheses wherein life emerged from more or less random chemical reactions when lightning combined compounds in the Earth's primitive atmosphere, forming organic molecules that rained into the sea. However, getting the molecules into sufficient concentration to form life's building

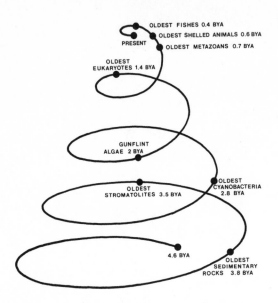

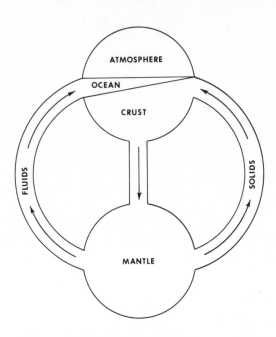

FIG. 3-1. Geologic time spiral beginning with the formation of the Earth 4.6 billion years ago (BYA).

FIG. 3-2. The steady-state model of the Earth.

blocks in such a short time after the Earth cooled would seem to be bordering on the miraculous.

An alternative theory presupposes that life started out in lumps of clay. In the laboratory, it can be shown that clays attract the antecedents of protein and DNA, the prime ingredients of life, and might have leached them from the sea during high tides. Furthermore, the clays trigger chemical reactions that string the building blocks into strands of protein and DNA. The clays can also store energy in the form of electrons, releasing them during periods of stress from the cycle of wetting and drying as the tides rise and recede. The energy could have triggered the chemical reactions that formed primitive proto-organisms.

Metals in the clay lattices also can form complexes with the precursors of proteins and DNA. The lattices might have given these building blocks a structure that was able to store energy, catalyze chemical reactions, and self-replicate: the most fundamental definition of life.

A place known today as North Pole in western Australia was once a tidal inlet, overshadowed by tall volcanic cones, erupting ash and lava that flowed

into the shallow sea. Thunderclouds hovered over the peaks, and lightning darted to and fro. Furious waves, whipped up by strong winds, pounded against the basaltic cliffs of the coastline. Further inland, the landscape was dominated by hummocks of black basalt flows, some still steamy from the latest eruptions. Everywhere, the air was pervaded with the rotten-egg smell of sulfur. Frequent downpours fed tidal streams that meandered onto a flat expanse of glistening, gray mud before reaching the sea. Elsewhere, there were scattered, shallow pools, containing highly saline water that periodically evaporated and left a variety of salts behind. Every so often, a flood tide washed across the mud flat, shifted the sediments, and replenished the brine pools.

North Pole is an ideal place to look for the origin of life. Although the climate is often extremely hot, the rocks of this area remained relatively cool throughout geologic history. This is an important fact because rocks that have been subjected to the intense heat of the Earth's interior have lost all traces of fossilized life forms. Unlike all the ancient rocks in the 3.5- to 3.8-billion-year range found throughout the world, only the rocks of the North Pole sequence have a history of low metamorphic temperatures. However, it is still difficult to prove that a mildly metamorphosed rock contains *microfossils*, which are preserved cell walls of once living unicellar microorganisms. Most of these fossils are simple spheres with few surface features and could just have easily been inorganic carbon compounds squeezed into spheroids by the growth of mineral grains deposited around them. However, some of these spheres were linked in pairs or chains and a few were in groups of four.

One indication that life possibly existed as far back as 3.5 billion years ago is structures called *stromatolites* (FIG. 3-3). They are less direct evidence of life since they are the remains not of the microorganisms themselves, but of sedimentary structures they built. Modern stromatolites are concentrically layered mounds of calcium carbonate. Bacteria or algae build them up by cementing sediment grains with a jellylike ooze. Ancient stromatolite colonies lived in the intertidal zones, and their height was indicative of the height of the tides. The oldest colonies were much taller than later ones, which supports the idea that the Moon was much closer to the Earth and caused higher tides at that time.

Fossil stromatolites are very similar to modern ones and are known to exist from as far back as 3 billion years. Older structures have been found at the North Pole and have been classified either as stromatolite fossils or as layered inorganic sedimentary structures. Microscopic filaments, radiating outward from a central point and resembling filamentous bacteria, have also been found, suggesting that the stromatolites were formed by bacteria.

Another way of pushing back the time scale of life is by chemical means. The biochemical reactions by which all living things harness energy for their growth give rise to distinctive ratios among the isotopes of certain elements. For instance, sulfate-consuming bacteria prefer the heavier sulfur-34 to sulfur-32. The oldest rocks found so far are the 3.8-billion-year-old Isua group in southwestern Greenland that show a depletion of carbon-13 with respect to the lighter carbon-12, a common

manifestation of biological activity.

In effect, life has a tendency to change the chemistry of the earth. The ratios of carbon isotopes in some early Precambrian rocks have been thought to indicate that photosynthesis was going on at that time. The early seas contained an abundance of iron, and any oxygen produced by photosynthesis was lost to the oxidation of iron. The abundance of sulfur in the early sea provided the nutrients to sustain life without the need for oxygen, and bacteria obtained energy by the reduction (opposite of oxidation) of sulfate ions.

Unfortunately, the chemical evidence is not totally reliable, and inorganic reactions also can give

(Photo by A.F. Shride, courtesy of USGS)

FIG. 3-3. Stromatolite structures.

rise to the same products as some biological reactions. This is no to say that the chemical data are not viable. When combined with other evidence such as the occurrence of microspheroids, filamentous microfossils, and stromatolites, they provide strong support for the existence of life much earlier than it was ever though possible.

THE PRECAMBRIAN POPULATION EXPLOSION

The earliest organisms were probably sulfur-eating bacteria somewhat similar to those living symbiotically in the insides of the tube worms that exist on the deep ocean floor near sulfurous hydrothermal vents (FIG. 3-4). Sulfur was particularly abundant in the early ocean, and the element combines easily with metals like iron to from sulfates. The atmosphere and ocean were totally devoid of oxygen, so the bacteria had to live under anaerobic condi-

tions, obtaining energy for their growth by the reduction of sulfate ions.

Even though this form of energy was quite satisfactory for the time being, the bacteria were letting a potential source of energy go to waste: sunlight. Around 2.8 billion years ago, microorganisms called *cyanobacteria* (formerly called blue-green algae) began to use sunlight as their main energy source to drive the chemical reactions needed to sustain their growth. These organisms lived within the top 200 feet of the ocean where the sunlight can penetrate easily, but they could not live on the surface of the water because the ozone layer, which filters out deadly ultraviolet radiation from the Sun, had not yet formed.

The development of photosynthesis was possibly the single most important feature in the evolution of life on Earth. It gave the cyanobacteria a practically unlimited source of energy. They could

(Courtesy of USGS)

FIG. 3-4. Clusters of tube worms and sulfide deposits on the ocean floor.

use sunlight to split water molecules and combine them with carbon dioxide to form simple sugars and proteins, liberating oxygen in the process.

However, oxygen is also poisonous to all life forms that have not evolved special defenses such as enzymes to protect against it. If the buildup of oxygen in the ocean remained unchecked, life certainly would have been in jeopardy. Fortunately, oxygen easily combines with metals to form metal oxides. Banded iron formations, an important source of iron ore, are scattered throughout the world and date between 2.6 and 2 billion years old. During this period, iron was precipitated out of the ocean in vast amounts.

These iron ore deposits attest to the fact that during the time of their formation, the ocean contained a significant amount of oxygen, although not nearly as much as it does today. It took 2 billion more years before the Earth's atmosphere contained nearly as much oxygen as it does today (FIG. 3-5).

Two billion years ago, the ocean probably had an upper oxygenated layer and a lower anaerobic layer that held vast amounts of iron. Mixing between the layers was accomplished by upwelling currents, similar to those operating on continental margins today. This mixing was not only important for the formation of iron ore deposits on continental shelves, but also was needed to replenish the upper ocean with nutrients such as phosphates and nitrates.

Before the end of the Precambrian era, when the oxygen level in the atmosphere was still less than 10 percent, organisms developed a method of obtaining energy by combining oxygen with nutrients in a process called respiration. Respiration freed the organisms from a dependence on sunlight, and the first simple animals made their appearance. At first there was no clear demarcation between animals and plants, and the organisms shared some characteristics of both, utilizing photosynthesis and respiration. As more mobile and complex forms of animals, called *metazoans*, evolved, they relied totally on respiration. The animals also fed upon plants and upon each other, creating a new predator-prey relationship.

With the depletion of carbon dioxide and other greenhouse gases in favor of oxygen, the Earth cooled sufficiently to bring on a major ice age, beginning some 2 billion years ago. The cooler climate effectively kept population growth down because the

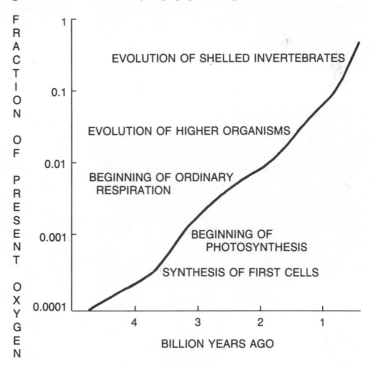

FIG. 3-5. The evolution of life and oxygen.

colder seawater temperatures kept populations narrowly confined to regions close to the equator.

Much of the life in the ocean today had ancestors in the Precambrian era, but large extinctions eliminated many species. The only record of their existence is fossils, which themselves only represent a fraction of all the life forms that ever existed on earth.

Since about 1.4 billion years ago, the previous skimpy fossil record of preserved cellular remains became much better. Another vast improvement came after another 700 million years had passed, thanks to tracks and body imprints made by soft-bodied, burrowing animals. The dominant species were coelenterates, jellyfishlike floaters up to 3 feet in diameter, and feathery forms more than a yard long and attached to the seafloor. The rest of the organisms were mostly marine worms, unusual naked anthropodlike animals, and a curious looking three-rayed, tiny naked starfish. A sheetlike marine worm grew nearly 3 feet long, but was less than $\frac{1}{10}$ inch thick.

As the Precambrian era came to a close, there was an explosion of animal life, and for the first time, hard external skeletons provided an excellent fossil record of an animal's existence. Animals no longer relied on exposed body surfaces to absorb oxygen directly from seawater, and gills and circulatory systems evolved when the oxygen level in the ocean reached about 10 percent of its present value.

GREENING OF THE EARTH

The drifting of the continents over the past 700 million years had a profound effect on the ocean currents, the temperature, the nature of seasonal fluctuations, the distribution of nutrients, the patterns of productivity, and many other factors of fundamental importance to living organisms. The evolution of marine animals varied throughout geologic time in response to these environmental changes. The vast majority of marine species lived on the continental shelves, in shallow water around islands, or on subsurface rises at depths less than 650 feet (FIG. 3-6). The richest shallow-water floras and faunas were in the warm tropics, where communities were packed with large numbers of highly specialized species each competing for a limited food supply. Every possible niche in the ocean was filled, but beyond the shoreline laid a vast untouched land just waiting to be conquered.

The Earth was over 4 billion years old and life had been around for nearly 3 billion of those years before land plants finally made their appearance about 400 million years ago. Prior to this time, plants lived in the shallow waters just below the surface, but one major factor kept them from venturing onto the land. The Sun's ultraviolet rays are harmful to plants, as well as animals, and as long as the Earth was bathed in this deadly light, plants were obliged to stay in the shelter of the water. By the time the oxygen level rose to a certain level, however, a tenuous layer of protective ozone formed in the upper atmosphere 20 to 30 miles above the Earth's surface.

The first land plants were probably algae and seaweedlike plants that lived in the intertidal zones and primitive forms of lichen and moss, which existed on exposed surfaces. They were followed by tiny, fernlike plants, which were the predecessors of trees. The fernlike plants lacked a root system and leaves, and were fertilized with spores attached to the ends of simple branches.

During their first 50 million years on dry land, plants displayed increased diversity and complexity in structure, including root systems, leaves, and reproductive organs using seeds instead of spores. A variety of branching patterns were developed to expose leaves to as much sunlight as possible and maximize photosynthesis. As plants become larger, they had to diverge from random branching to tiers of branches in order to achieve greater efficiency, which also required stronger structural support. These innovations gave rise to the preponderance of flora we have today.

Beginning about 700 million years ago, marine animals developed an internal skeleton, one of the most significant achievements in life's long climb up the evolutionary ladder. Invertebrates, using an external skeleton for the support of their bodies, were at a distinct disadvantage when it came to mobility and growth. Vertebrates, on the other hand, have an internal skeleton that is light, strong, and flexible, and has more efficient attachments for muscles. The skeleton grows as the animal grows.

Internal skeletons allowed species to become free-swimming and widely dispersed into a variety of environments. The first fish were small mud-grubbers and sea squirts, which lacked jaws and teeth. Despite their obscure beginnings, however, fish quickly became the dominant species of the ocean, and the sharks became masters of the deep.

The first land animals did not put in their appearance until the plants were well established some 350 million years ago. These were the first primitive amphibians: fish with lobe fins modified into walking limbs, and lungs or gills adapted for breathing air. What might have enticed these creatures to come ashore was an abundance of food swept up on the beaches during high tide. Fierce competition in the ocean for a scarce food supply opened the field for any animal that could find food on land. The early amphibians probably did not stay on shore for very long at a time because their primitive legs could not support their body weight for long periods. Eventually, as their limbs strengthened from digging in the sand for food and shelter, some amphibians were encouraged to wander farther inland, where crustaceans and insects were abundant. The amphibians soon dominated the land and were particularly attracted to the great swamps from which came most of our coal deposits. When the swamps began to dry out, the amphibians gave way to their cousins, the

(Courtesy of NOAA)

FIG. 3-6. A diver exploring the seafloor.

reptiles, which were more suited to a continuous life on dry land.

During the Mesozoic era, or Age of Middle Life, from 240 to 65 million years ago, life in the ocean saw many remarkable advancements. Those species that made it through the great extinctions at the end of the Paleozoic era were similar to populations found today. Many regions became filled with numerous specialized animals, and the overall diversity of species in the world's oceans rose to unprecedented heights. Species that were stationary and anchored to the ocean floor generally declined in relation to their more mobile counterparts.

It was also during this era that the breakup of the supercontinent of Pangaea began, and continents started to migrate to their present positions. Land plants bore little resemblance to those of the previous era and were more closely related to those of the present. The Mesozoic era is also known as the ''age of the dinosaurs,'' and dinosaurs dominated every continent. Less dramatic were the mammals which, during this period, were little more than small rodentlike creatures. They would eventually take their place as the dominant land species when the dinosaurs became extinct at the end of the Mesozoic era.

ALIENS ON THE DEEP-SEA FLOOR

On the crest of the East Pacific Rise is a world that time forgot. The East Pacific Rise is a 6000-mile-long rift system along the eastern edge of the Pacific plate and is the counterpart of the Mid-Atlantic ridge.

Using deep-sea submersibles, scientists have been able to explore this undersea mountain range 2½ miles beneath the ocean's surface. At the base of jagged basalt cliffs is evidence of active lava flows and fields strewn with pillow lava. There are also active hydrothermal fields where seawater seeps down to the magma chamber, is heated, and is expelled through vents, forming underwater geysers (FIG. 3-7). The hot water is rich in dissolved minerals that precipitate out, building exotic-looking chimneys. Some chimneys spew out water blackened with sulfide minerals and are called for obvi-

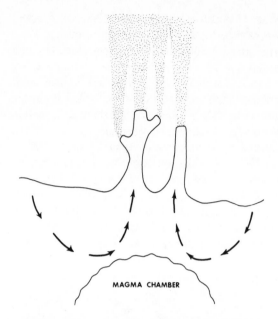

FIG. 3-7. Hydrothermal vents on the ocean floor.

ous reasons *black smokers* (FIG. 3-8). Others expel water that is milky white, and for this reason, they are called *white smokers*.

With further exploration of this new bizarre world, scientists made another astonishing discovery of great import. Flourishing among the hydrothermal vents, live what are perhaps the strangest animals ever found on earth. Large white clams up to a foot long were nestled between black pillow lava. Giant white crabs, having no need for skin pigments, scampered blindly across the volcanic terrain. They live in total darkness and therefore are sightless, having no need for eyes.

Most dramatic of all for the scientists were clusters of giant tube worms, some as tall as 10 feet, weaving eerily in the hydrothermal currents. Attached to the upper ends of the long, white stalks are bright red plumes, resulting from the presence of oxygenated hemoglobin in their blood. Apparently, the plumes are a delicacy to the crabs, which climb the stalks in order to find a meal.

Normally, water this deep is near freezing, but the hydrothermal vents keep the water at a balmy temperature, upwards of 70 degrees Fahrenheit in some places. Because sunlight cannot penetrate to these great depths, the communities are entirely independent of solar energy, which is required by all other living organisms on Earth. Instead, they rely on the flow of energy from the Earth's interior.

In addition to warmth, the black smokers also provide nutrients mostly hydrogen sulfide. At the base of the food chain are bacteria that oxidize hydrogen sulfide into elemental sulfur and various sulfates. The bacteria harness the energy liberated by the oxidation in order to incorporate carbon dioxide into organic matter to make carbohydrates, proteins, and lipids.

Most of the animals either feed on the bacteria directly or live with them symbiotically. The bacteria live in the gills of clams and mussels or in a special organ inside the tube worms, and the byproducts of the bacteria's metabolism leak into the host animal and nourish it. The vent animals are so dependent on the sulfide-metabolizing bacteria, that the mussels have only a rudimentary stomach and the tube worms lack even a mouth. The animals live precarious lives since the hydrothermal vent systems turn on and off in sporadic cycles. Indeed, iso-lated piles of empty clamshells bear witness to local mass fatalities.

EXTINCTIONS ARE A WAY OF LIFE

It is a commonly held belief among scientists that environmental change drives evolution. In 1979, the British chemist James Lovelock proposed his Gaia hypothesis, which presupposes that life can control its environment. In other words, living organisms maintain the optimum conditions for life by regulating the climate, like the human body regulates its own temperature. There is also a suggestion that from the very beginning, life followed a well-organized pattern of growth, rather than a haphazard one. Life kept pace with all the changes in the Earth over time and even made some changes of its own, such as converting most of the carbon dioxide in the atmosphere into oxygen.

Yet the fossil record seems to indicate that life evolved by fits and starts. Evolution, therefore, was not gradual and of constant tempo, but instead, there were short periods of rapid change separated by long periods of little change. Evolution might also be op-

(Courtesy of USGS)

Fig. 3-8. Black smoker on the East Pacific Rise.

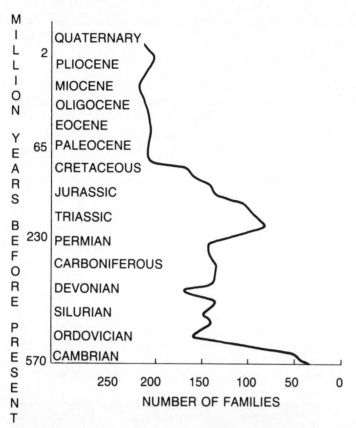

FIG. 3-9. The number of families through time.

portunistic, however, in that variations arise by chance and are selected in accordance with the demands of the environment. When the environment changes abruptly to one that is not benevolent, species that cannot adapt to these new conditions cannot live at their optimum nor pass on their "bad" genes to future generations.

It was hard medicine to take, but geologists are finally coming around to accept catastrophe as a normal part of Earth history and therefore, as a part of the uniformitarian process, also known as gradualism. More scientists are accepting the impact theory of extinction, which maintains that the bombardment of one or more large cosmic invaders brought on periodic mass extinctions. Therefore, mass extinctions appear to be part of a pattern of life throughout the Phanerozoic eon, the past 700 million years. Each extinction resets the evolutionary clock, making the history of life on Earth spasmodic and governed by happenstance. Since the great Per-

mian die out 240 million years ago, there have been five major and five minor extinctions, occurring about once every 26 million years (FIG. 3-9).

A classic example is the extinction of the dinosaurs and 70 percent of all other known species at the end of the Cretaceous period, 65 million years ago. A thin layer of mud composed of shock-impact sediments exists at the Cretaceous-Tertiary rock boundary in many parts of the world. Within this layer is a high concentration of iridium, an isotope of platinum that is extremely rare on Earth but relatively abundant on meteors, asteroids, and comets. The iridium is a thousand times greater than normal background concentrations, indicating that impacts from one or more cosmic visitors cast up tremendous amounts of dust in the atmosphere. This dust might have lingered for months and shaded the Earth. It also might have cooled the climate sufficiently to cause the extinctions of many species.

One interesting theory to account for the periodicity of the extinctions presupposes that the Sun has a small, dark companion star called Nemesis that orbits close by the Sun every 26 million years and distorts the Oort Cloud of orbiting comets, which then rain down upon the Earth.

Today, we live in a highly diverse world with as many species that have ever lived at any other time, associated with a rich variety of communities, living in a large number of provinces, possibly the richest and largest that ever existed. We have been furnished with a biosphere that is highly diverse and most interesting.

4

The Global
Garbage Disposal

BY the turn of the twentieth century, scientists pondered why the continents seemed to fit together like a giant jigsaw puzzle, why the rocks of mountain ranges on opposite shores of continents were the same, why the past climatic conditions were so much similar, and why there were identical plant and animal life on continents now widely separated by oceans. In 1915, the German meteorologist and arctic explorer Alfred Wegener suggested that the only reasonable explanation was that the continents were once together in one large continent he named Pangaea. Most scientists of his day called his theory preposterous, and he was made an outcast by the scientific community. Alfred Wegener's theory of continental drift was not accepted until 30 years after his death, which occurred during an expedition to Greenland. Incorporated into a new theory of plate tectonics, the drifting of the continents neatly explained many of the unsolved problems that have plagued scientists for decades. It also has made them more aware that the Earth is a dynamic planet, immensely more complicated than it was previously thought.

MYSTERIOUS MOUNTAIN RANGES

North America, South America, Greenland, Europe, and Africa formed one expanse of land 165 million years ago. When these continents are fitted together at their present continental shelves (FIG. 4-1), which would be the point of rupture, it is remarkable how well they have retained their original shapes, although some distortion is inevitable, considering what the continents have gone through over the past millions of years. If they could be brought together again today, Greenland would fit neatly into the coastline of northern Europe, North America would hug West Africa, and the bulge of South America would fit tightly into the cleft of Africa, although there would have to be some slight overlap to allow for any distortions.

By 125 million years ago, the infant North Atlantic Ocean, created by the rift between North America, Eurasia, and Africa, had an active midocean ridge that began to make new oceanic crust. At about the same time, South America began to separate from Africa and moved with Africa away from North America and Europe. About 80 million

FIG. 4-1. The fit of the continents.

years ago, the North Atlantic became a full-fledged ocean, and North America and Greenland began to drift apart. At that time, the South Atlantic began to develop, opening up from south to north like a zipper. The creation of a new ocean basin also dropped the sea level by about 100 feet, leaving many parts of the continents that were previously flooded high and dry. About 20 million years ago, a ridge near Iceland subsided, allowing cold seawater from the recently formed Arctic Ocean to surge into the Atlantic, giving rise to the vigorous circulation the ocean has now.

The ocean floor was once thought to be a barren and featureless landscape covered with thick, muddy sediments washed off the continents. During the mid-1800s, lead soundings made in preparation for laying the first transcontinental telegraph cable linking the United States with Europe told of

hills and valleys and a mid-Atlantic rise, named Telegraph Plateau, where the ocean was supposed to be the deepest.

In 1874, the British cable-laying steamship H.M.S. *Faraday* was sailing in the North Atlantic on a mission to mend a broken telegraph cable. The cable had broken in water 2 1/2 miles deep where it passed over a large rise in the ocean floor. While grappling for the broken cable, the ship caught the strong claws of the grapnel on a rock. With the winch straining with all its might to free the grapnel, it finally gave way and was brought to the surface. Clutched in one of its claws was a large chunk of black basalt, a volcanic rock found in what was thought to be the most unlikely place for a volcano.

The first fully equipped oceanographic vessel was the British corvette H.M.S. *Challenger*, which was commissioned to explore the world's oceans in

1872. The crew took soundings, water samples, and temperature readings, and dredged bottom sediments for samples of animal life living on the deep ocean floor. They discovered hundreds of species that had never been seen before. After nearly 4 years, the *Challenger* charted 140 square miles of ocean bottom and sounded every ocean except the Arctic. Their deepest sounding was at a depth of 5 miles off the Marianas Islands in the Pacific trench. No other ocean exploration of this magnitude was ever attempted again until after the world wars.

The invention of sonar in World War I used to detect submarines by bouncing sound waves off them, gave scientists an important tool for mapping the ocean floor. By World War II, most ships had a *fathometer*, a special type of sonar that can determine the depth of the ocean floor and record it on a strip chart recorder. As ships criss-crossed the Atlantic Ocean during the war, their echo sounders painted a picture of the ocean floor never before imagined (FIG. 4-2).

Showing up on the sonars was a huge submarine mountain range, running through the middle of the Atlantic. The range surpassed in scale the Alps and Himalayas combined. What was even more remarkable was that it bisects the Atlantic almost exactly down the middle, weaving halfway between continents, and assuming the shapes of continental margins on opposite shorelines (FIG. 4-3).

As more detailed maps were made, the Mid-Atlantic Ridge, as it was called, became the most peculiar mountain range ever seen. Down the middle of the 10,000-foot-high ridge crest runs a deep trough, as though it were a giant crack in the Earth's

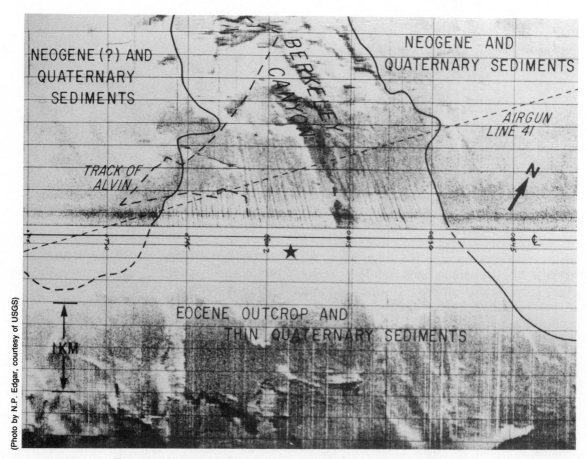

(Photo by N.P. Edgar, courtesy of USGS)

FIG. 4-2. Sonograph of the lower continental slope off the Atlantic coast.

FIG. 4-3. The Mid-Atlantic Ridge.

crust. It is as deep as 4 miles, or four times deeper than the Grand Canyon, and 15 miles wide, making it the longest and deepest canyon on the planet.

The Mid-Atlantic Ridge was later found to be part of a worldwide midocean ridge system, stretching over 40,000 miles and encircling the globe like the stitching on a baseball. The axis of the midocean ridge is offset laterally in a roughly east-west direction (FIG. 4-4). The offsets range from a few miles to a few hundred miles and are encountered every 20 to 60 miles along the length of the ridge.

The offsets are *transform faults*, where pieces of oceanic crust slide past each other. The friction between the plates gives rise to strong shearing forces that wrench the ocean floor into steep canyons. Transform faults are seismically active because the grinding of the plates along them generates strong earthquakes. The crest of the midocean ridge is also a center of intense seismic and volcanic activity, and the focus of a great flow of heat from the interior of the earth.

In the early days of sampling the sediments on the sea floor, scientists used a *dredge*, a sort of bucket on the end of a line. The major problem with

this technique was that it could sample only the uppermost layers and could not recover them in the order they were laid down. During World War II, neutral Sweden conducted expeditions to attempt to return to the surface samples of the vertical section through the ocean floor. Swedish scientists invented a piston corer, consisting of a long barrow that fell through the mud under its own weight and a piston that was fired upward from the bottom of the barrow, sucking up sediments into the pipe (FIG. 4-5). What the Swedish scientists brought to the surface were long, cylindrical cores of the ocean bottom dating back millions of years.

In 1968, the oceanographic research vessel *Glomar Challenger* was commissioned for the Deep Sea Drilling Project, a consortium of American oceanographic institutions. Its primary purpose was to take rotary core samples of the ocean floor at hundreds of sites in scattered parts of the world. Looking like an ocean-going oil-drilling derrick, the ship dangled beneath it a string of drill pipe up to 4 miles long. When the drill bit reached the ocean bottom, it bored through the sediments under its own weight. The core was retrieved through the drill stem and

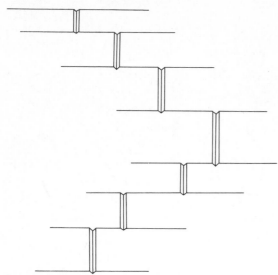

FIG. 4-4. Transform faults at the midocean ridges.

brought to the surface for scientists to study.

What the scientists found was truly remarkable. After dating several cores taken from around the midocean ridges, they found that the sediments were older and thicker the farther the ship drilled away from the ridge crests. What was even more remarkable was that the thickest and oldest sediments were found not to be billions of years old, but less than 200 million years old.

The ocean floor was once thought to contain sediments washed off the continents and debris from dead marine organisms, which were several miles thick from layer upon layer of accumulation. In order to measure these sediments, scientists invented a seismic device for use underwater. On land, seismic waves, which are similar to sound waves, are used to locate sedimentary structures that could trap oil. At sea, an explosive charge or an air gun was set off on the ocean floor, and because seismic waves travel slower in soft sediments and faster in hard rock, scientists could calculate the thickness of different rock layers. Another instrument they used was an ocean-bottom seismograph, which was lowered to the ocean floor. It recorded microearthquakes in the oceanic crust and automatically rose to the surface to be recovered.

These geophysical methods have given scientists information about the ocean floor that could not be obtained directly. Some of their findings came as a complete surprise, however. Instead of miles of sediment, they found an average thickness of only a couple thousand feet. Perhaps there was some sort of natural vacuum cleaner that swept the sediments off the ocean floor.

The sediments that end up on the ocean floor are detritus and the shells and skeletons of dead microscopic organisms, which flourish and die in the sunlit waters of the top 300 feet of the ocean. Detritus, whose source is the weathering of rocks on land along with decaying vegetable matter from land plants is carried by rivers to the edge of the continent and out onto the continental shelf where the material is picked up by marine currents. When the detritus reaches the edge of the shelf, it falls to the base of the continental rise under the pull of gravity. Also, a significant amount of terrestrial material is blown out to sea by strong desert winds in subtropical regions. Approximately 15 billion tons of continental material reaches the outlets of rivers and streams annually. Most of the detritus is trapped near the outlets and on continental shelves, and only a few billion tons actually escapes into the deep sea.

The biological material in the sea contributes about 3 billion tons of sediment on the ocean floor each year. The rates of accumulation are governed by the rates of biological productivity, which are controlled in large part by the ocean currents. Nutrient-rich water upwells from the ocean depths to the sunlit zone, where the nutrients are incorporated into the cells of microorganisms. Areas of high productivity and high rates of accumulation are normally around major oceanic fronts, such as the region around Antarctica and along the edges of major currents such as the Gulf Stream and the *Kuroshio*, or Japan, current that circles clockwise around the North Pacific basin. Nutrient-rich water also upwells in a zone along the equator, such as along the coasts of Ecuador and Peru, and supports a major fishing industry there.

The rate of marine-life sedimentation is influenced by the ocean depth. The farther the shells must descend, the lesser are their chances of reaching the bottom before becoming dissolved. A shell's survivability also depends on how quickly it is cov-

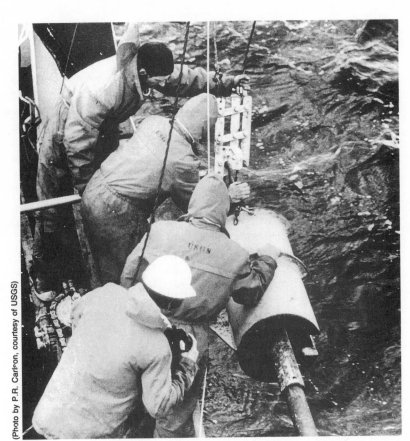

FIG. 4-5. Piston coring in the Gulf of Alaska.

(Photo by P.R. Carlson, courtesy of USGS)

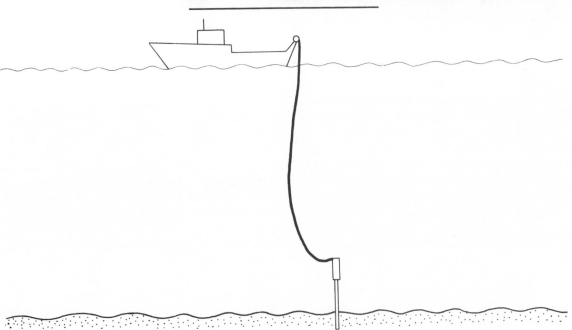

ered up and protected from the corrosive action of the deep-sea water.

If there were no bottom currents and only marine-born sedimentation, there would be an even blanket of material settling onto the original volcanic floor of the world's oceans. As it is, however, the rivers of the world also contribute a substantial amount of the material that ends up on the deep ocean floor.

The largest rivers of North and South America empty into the Atlantic, which receives considerably more river-borne sediment than does the Pacific. The Atlantic is smaller and shallower than the Pacific, so its marine sediments are buried more rapidly and therefore are more likely to survive than their Pacific counterparts. This process might also have greatly aided the formation of offshore petroleum reservoirs. Moreover, the deep-ocean trenches around the Pacific trap much of the material that reaches its western edge, where it is subducted into the mantle. Thus, on the average, the floor of the Atlantic receives considerably more sediment (about 1 inch per 2500 years) than the floor of the Pacific. In addition, strong near-bottom currents redistribute sediments in the Atlantic on a grander scale than they do in the Pacific.

THE SPREADING SEAFLOOR

One of the most surprising results of the Deep Sea Drilling Project was the discovery of thin layers of calcium-carbonate sediments covered by a thick sequence of red clay. These layers were taken from cores of the abyssal plains near the continental shelves. The deepest abysses are found adjacent to the continental margins, where the oceanic lithosphere is the oldest. At the calcium-carbonate compensation zone below a depth of about 3 miles, calcium carbonate derived from the shells of microorganisms dissolves in seawater because it is undersaturated in calcium carbonate, whose solubility increases with pressure. However, the calcium-carbonate sediment layer was found about 4 miles below the ocean surface and well protected from seawater by the overlying mud. The only plausible explanation was that the calcium-carbonate layer had to form in the shallower water near the midocean

ridge and somehow made its way to the edges of continents.

The depth at which the oceanic crust sinks as it moves away from the midocean ridges varies with the square root of its age; therefore, crust that is 2 million years old lies about 2 miles deep; crust that is 20 million years old lies about 2.5 miles deep; and crust that is 50 million years old lies about 3 miles deep. The oceanic crust is composed of an upper layer of pillow basalts, which was formed when lava was cooled by seawater; a middle layer composed of a sheeted-dike complex, which consists of a tangled mass of feeders that brought the lava toward the surface; and a lower layer which is composed of *gabbros*, coarse-grained basalts that crystallized slowly under high pressure.

The ocean floor at the crest of the midocean ridge consists almost entirely of hard volcanic rock. With increasing distance from the crest, the rock is covered by an increasing thickness of soft sediment, mostly calcium carbonate and red clay, with calcium carbonate predominant nearer the ridge crest. This implicates the midocean ridges as the source of the calcium carbonate layers.

Over the Earth's long history, the magnetic poles have reversed themselves. The Earth's magnetic field reverses in irregular intervals approximately five times every million years for unknown reasons. Volcanic rocks have a high concentration of iron, and as the basalt cools, it passes through the Currie point, at which temperature the iron molecules line up with the Earth's magnetic field, as though they were miniature bar magnets.

Magnetometers were towed over the midocean ridges and indicated stripes of reversed magnetic fields extending outward and parallel to the ridge crests. What is most puzzling was that the magnetic stripes are mirror images of each other on both sides of the midocean ridge (FIG. 4-6). It appears that each succession of basaltic flows pushed the latter flow farther away from the ridge crest. As the basalt cooled, its iron molecules were polarized in the same direction as the Earth's magnetic field. Normal polarities in the rocks were reinforced by the present magnetic field, and reversed polarities were weakened by it. The only reasonable explanation for

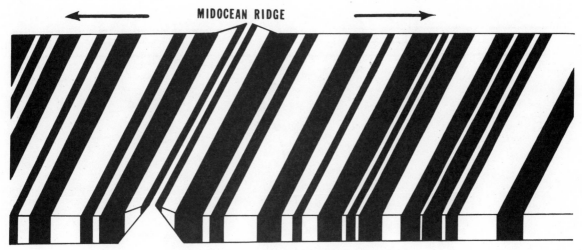

FIG. 4-6. Magnetic stripes on the ocean floor.

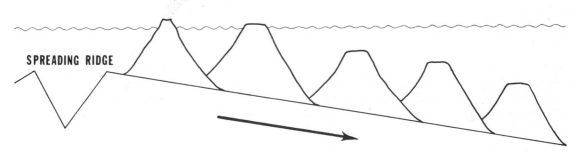

FIG. 4-7. Guyots on the ocean floor.

the identical parallel bands of magnetic rocks on both sides of the midocean ridge is that the ocean floor must spread outward from the ridge crests. Age-dating the stripes not only indicated the age of the ocean floor, but also the rate that the seafloor was spreading apart. For the Atlantic this rate is about 1 inch per year.

In the Pacific Ocean, there are strange-looking, extinct undersea volcanoes called *guyots* (FIG. 4-7). They have flat tops as though the peaks of the cones were sawed off. The volcanoes were at one time above sea level, and through time, wave erosion gradually wore them down below the surface of the sea. What is even more interesting is that the farther away the guyots are from the volcanically active regions of the ocean, the older and squatter they

are. This seems to indicate that the guyots wandered across the ocean floor, away from their places of birth. The Hawaiian Islands are a chain of volcanoes. The island of Hawaii is the youngest and most volcanically active and lies to the southeast. The rest of the islands are in a line to the northwest, with each being older than the one to its southeast. It appears that the islands were produced assembly-line fashion, with each one moving away in succession from its point of origin as though on a conveyer belt.

About 100 years ago, many scientists thought the Earth was expanding. Earlier theories that explained how the Earth maintained its internal temperature hot enough to melt rock suggested that it was shrinking by gravitational collapse. The discov-

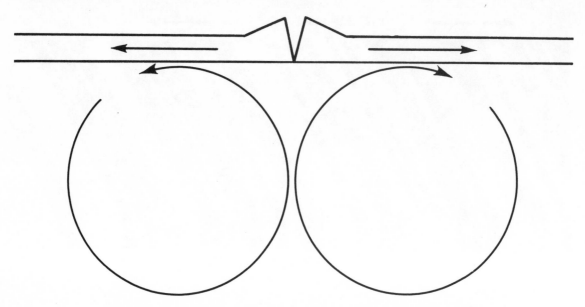

FIG. 4-8. Convection currents in the mantle spread lithospheric plates apart.

ery of radioactivity in the late 1880s provided scientists with a theory for an alternative source of energy. Radioactive decay of certain unstable elements heated up the Earth's interior, and instead of the Earth shrinking, scientists reasoned that the Earth was expanding in order to get rid of the excess heat. Also, the weakening of the Earth's gravitational field would cause it to bulge, forming cracks in the crust like those on a boiled egg.

The expanding-Earth hypothesis could easily account for the separation of continents as though they were painted on a balloon and inflated. The problem with this theory is that if the Earth today were significantly larger than it was in the past, there would be obvious defects in the shapes of the continents, and they would not fit together as well as they do.

Instead, the American geologist Harry Hess proposed an expansion of a different sort. Hot rock heated from below rose to the surface of the mantle, bled out at the midocean ridges, and pushed the oceanic crust apart in a process he called *seafloor spreading*.

As the rocks heat up in the *asthenosphere*, the partially molten layer of the upper mantle, they become plastic, slowly rise by convection, and after

millions of years, reach the topmost layer of the mantle, the lithosphere. When the rising rocks reach the underside of the lithosphere, they spread out laterally, cool, and descend back into the interior. The constant pressure against the bottom of the lithosphere causes fractures to form. As the convection currents flow out on either side of the crack, they carry the two separated parts of the lithosphere along with them, and the fracture widens (FIG. 4-8).

With reduced pressure, the rocks melt and rise up through the crack, where the molten rock, or magma, finds easy passage through the 60 miles or so of lithosphere until it reaches the oceanic crust. There, it forms magma chambers that further press sideways against the oceanic crust, widening it while at the same time providing molten lava, which pours out from the trough between the two ridge crests, adding layer upon layer to both sides of the spreading ridge. The surging magma also forms volcanic piles, some of which rise to sea level and become volcanic islands. On Iceland, which straddles the Mid-Atlantic Ridge, the spreading ridge system actually comes ashore, dissecting the country in two.

The pressure of the upwelling magma forces the ridge farther outward on both sides, pushing the ocean floor and the lithosphere upon which it rides

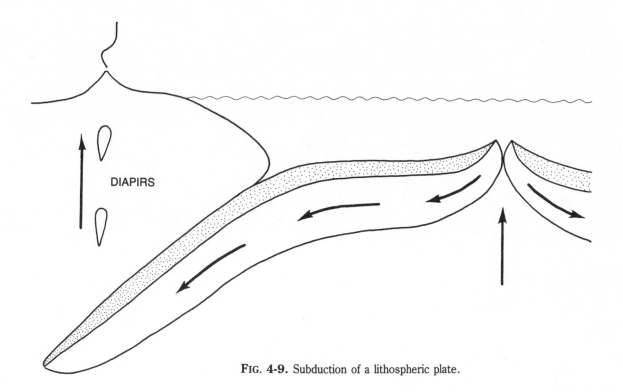

FIG. 4-9. Subduction of a lithospheric plate.

away from the mid-ocean ridge. As new material is added to the ocean floor, however, it must be subtracted somewhere else.

SWALLOWING THE OCEAN FLOOR

In order to conserve mass on the Earth's surface, the creation of new oceanic crust at the midocean ridges must be equally matched by the destruction of old oceanic crust. This swallowing up of the oceanic crust is accomplished in the deep-sea trenches, called *subduction zones*, where the crust is pushed down into the mantle (FIG. 4-9). Most of these trenches are in the western Pacific Ocean. As the rigid lithospheric plate carrying the oceanic crust descends into the mantle, it slowly breaks up, and over millions of years, it is absorbed into the general circulation of the mantle.

The subduction of the lithosphere plays the most significant role in global tectonics and accounts for many of the geologic processes that shape the surface of the Earth. The seaward boundaries of subduction zones are marked by the deepest trenches

in the world and are usually at the edges of continents or along volcanic island chains (FIG. 4-10). Major mountain ranges and most volcanoes and earthquakes are associated with the subduction of lithospheric plates, and the earthquakes act as beacons, marking the boundaries of the plates.

The amount of subducted plate material is vast, but keep in mind that when the Atlantic and Indian oceans opened up and new oceanic crust was created, an equal area had to disappear. This means that 5 billion cubic miles of crustal and lithospheric material was destroyed, and that at the present rate of subduction, an area equal to the entire surface of the Earth will be consumed by the mantle in the next 160 million years.

Where two lithospheric plates converge, it is usually the oceanic plate that is bent and pushed under the thicker and more stable continental plate (FIG. 4-11). The line of initial subduction is marked by a deep-ocean trench. At first the angle of descent is low, and then it gradually becomes steeper to an angle of about 45 degrees. At this angle, the rate

FIG. 4-10. The major trenches of the world.

TABLE 4-1. Dimension of Deep Ocean Trenches.

TRENCH	DEPTH (MILES)	WIDTH (MILES)	LENGTH (MILES)
Aleutian	4.8	31	2300
Japan	5.2	62	500
Java	4.7	50	2800
Kuril-Kamchatka	6.5	74	1400
Marianas	6.8	43	1600
Middle America	4.2	25	1700
Peru-Chile	5.0	62	3700
Philippine	6.5	37	870
Puerto Rico	5.2	74	960
South Sandwich	5.2	56	900
Tonga	6.7	34	870

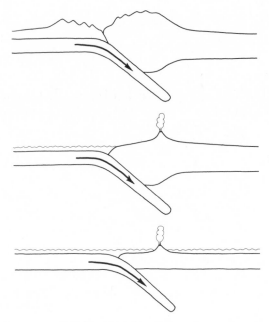

Fig. 4-11. Collision between plates. Two continental plates (top); an oceanic and continental plate (middle); two oceanic plates (bottom).

of vertical descent is less than that of the horizontal motion of the plate, and is typically 2 to 3 inches per year.

Heat flows into the cooler lithosphere from the surrounding hot mantle. The conductivity of the rock increases with increasing temperature, and the conductive heating becomes more efficient with depth. Heat of compression is introduced into the lithospheric plate as it descends and is subjected to increasing pressure. Heat is also generated within the lithospheric plate by the decay of radioactive elements, the change in mineral structure of the rocks, and internal and external friction, especially at the boundaries between the moving plate and the surrounding mantle. Among these heat sources, conductive heating and friction contribute the most toward the heating of the descending lithosphere.

At first, the interior of the plate remains relatively cool compared to the outside. As the plate penetrates to a depth of about 375 miles, its interior begins to heat up more rapidly because of the more efficient transfer of heat by radiation. When the plate reaches a depth of 430 miles—the boundary between the upper and lower mantle—it ceases to be thermally distinguishable as a structural unit from the mantle; in effect, it becomes a part of the mantle.

Deep earthquakes are associated with descending plates, and the cessation of earthquakes at 430 miles seems to indicate that the plate either has stopped descending or that it has become too hot to generate earthquakes.

Not all plates descend even to this level, however. The depth the plate reaches before being assimilated by the mantle depends on its rate of descent. A slowly moving plate will attain thermal equilibrium at shallower depths. For example, if the plate were descending at a rate of one-half inch per year, it would only travel to a depth of about 250 miles before complete assimilation occurred. If a quickly descending plate should reach the boundary between the upper and lower mantle, it would be bent horizontally and move parallel to that boundary until it was completely assimilated into the mantle.

When continental crust moves into a subduction zone, its buoyancy prevents it from being carried farther than 25 miles below its normal depth. When two continental plates converge, the crust is scraped off the subducting lithospheric plate and plastered onto the overriding plate, while the lithospheric plate without its crust continues its dive into the mantle. The submerged crust is underthrusted by more crust, and the increased buoyancy pushes up mountain ranges such as the Alps and Himalayas. Further compression and deformation might take place farther inland beyond the line of collision, producing a high plateau with surface volcanoes, like the plateau of Tibet in China. As resisting forces continue to build up, the plate convergence eventually will stop.

Continental collisions such as when India broke off of Gondwanaland and slammed into Asia, forming the Himalayas, also might be responsible for periodic reorientations of the plates and might explain why Pacific undersea volcanoes, like the Emperor seamounts northwest of the Hawaiian islands, made an abrupt turn to the north about the time the North American plate rammed into the Pacific plate.

FIG. 4-12. The carbon cycle.

The deep-ocean trenches, created by descending lithospheric plates, accumulate large deposits of sediments from the adjacent continent. The continental shelf and slope, consisting of sediments washed off the continents and the remains of dead marine life from above, might extend several hundred miles from the edge of the continent. As these sediments and their content of seawater are caught between a subducting oceanic plate and an overriding continental plate, they are subjected to strong deformation, shearing, heating, and metamorphism. Some of the sediments might be carried deep within the mantle, where they are melted in pockets called *diapirs*. Like huge bubbles, the diapirs rise to the base of the continental crust, where they become the source of new magma for volcanoes and granitic rocks. In this manner, the continental crust is constantly being rejuvenated, and the total mass of low-density crustal rocks is always preserved.

THE GREAT RECYCLING PLANT

The oceans hold the largest store of carbon dioxide on Earth—as much as 60 times greater than that in the atmosphere, and mostly in the form of dissolved bicarbonate. The carbon dioxide dissolves in seawater much like it does in soft drinks to give them their fizz. Carbon dioxide enters the ocean by surface wave action, and the concentration of carbon dioxide in the topmost 250 feet is as much as that in the entire atmosphere.

In this mixed layer of the ocean, microorganisms convert the carbon dioxide into calcium carbonate, which forms a calcite ooze on the ocean floor and eventually hardens into limestone. If this material should fall into the deep sea, it is dissolved in the cold waters of the abyssal. The abyssal by virtue of its great volume, holds the vast majority of free carbon dioxide, or that which is not locked up in carbonaceous sediments.

The capacity of the abyssal to absorb carbon dioxide is almost limitless. Nevertheless, carbon dioxide moves from the atmosphere, through the mixed layer of the ocean, and into the oceanic depths very slowly and at a nearly constant rate. Unfortunately, this is only half the rate it is being released by man's activities, which includes the burning of fossil fuels and forests.

Most of the world's carbon is stored in the biosphere and in sediments on the continents and on the ocean floor (FIG. 4-12). The amount of carbon in the form of carbon dioxide in the original atmosphere was thousands of times greater than it is today. As the continents grew, however, they took carbon out of the oceans and the atmosphere, and locked it up in carbonaceous sediments.

Carbon dioxide is rained out in the form of a weak carbonic acid, which leaches minerals out of rocks, especially calcium. Streams transport the minerals to the ocean. In the ocean, the calcium minerals are taken up in the shells of marine organisms. When the organisms die, their shells sink to the bottom of the sea, where they slowly build up deposits of limestone. If this scrubbing of carbon dioxide out of the atmosphere and storing it in limestone continued unchecked, the atmosphere would be depleted of carbon dioxide. Without this important greenhouse gas, the climate would turn cold and possibly bring on a new ice age.

It was not until the development of the theory of plate tectonics, spreading ridges, and subduction zones on the ocean floor that the mystery of the missing carbon dioxide was solved. The ocean crust is relatively young, less than 5 percent of the Earth's age. The ocean floor is continually being created at the midocean ridges and destroyed in the trenches. When the seafloor is forced into the Earth's interior, carbon dioxide is driven out of the limestone by the intense heat. The carbon dioxide and water trapped in the sediments become the dominant volatiles in the fluid melt, which works its way through the mantle and eventually into the roots of volcanoes or midocean ridges. The eruption of volcanoes and the flow of molten basalt from the midocean ridges resupplies the atmosphere and the ocean with new carbon dioxide and water, making the Earth one great chemical-recycling plant.

Carbon dioxide plays an important role in regulating the temperature of Earth, and any changes in the carbon cycle will have profound effects on the climate. As the early Sun heated up and temperatures on Earth rose, more water was evaporated from the oceans, which increased rainfall on the land. This speeded up the weathering process, and most of the chemical weathering at the surface of the Earth resulted in the removal of atmospheric carbon dioxide, which was converted to limestone on the seafloor. The drop in levels of carbon dioxide in the atmosphere kept the Earth from overheating. If the Earth's temperature began to get too cool, less water was evaporated from the ocean, chemical and biological reactions became sluggish, and less carbon dioxide was removed from the atmosphere, even though the input from volcanoes and midocean ridges remain somewhat constant. Thus, the carbon cycle is also Earth's thermostat, keeping the planet's temperature within tolerable limits for life.

5

Volcanoes of the Deep

VOLCANOES replenish the Earth with the vital ingredients it needs to stay alive. These ingredients include important gases, such as carbon dioxide, nitrogen, and water vapor, along with volcanic rocks from which to build new crust. More than 80 percent of the Earth's surface above and below sea level is of volcanic origin. Most of the world's islands started out as undersea volcanoes, and successive eruptions piled up volcanic rocks until the volcano finally broke the water's surface (FIG. 5-1).

Mauna Kea and Mauna Loa are the two principal volcanoes that created the island of Hawaii. In reality, Mauna Kea is the tallest mountain in the world, for it rises some 32,000 feet above the ocean floor, making it several hundred feet taller than Mount Everest.

Oceanic volcanoes also happen to be among the most explosive volcanoes in the world (FIG. 5-2), and whole islands have been known to disappear when a volcano blew off its top. New volcanic islands are being born as well, however. In only a few years after they have made their appearance, life takes root in the rich volcanic soil, and the island is on its way to becoming a lush paradise.

A DIPLOMAT BLAMES
THE WEATHER ON A VOLCANO

Iceland occupies a unique position in the world. As its name indicates, Iceland is one of the coldest inhabited places on Earth. However, Iceland is also one of the most thermally active countries in the world and is blessed with an abundant geothermal energy source for generating electricity and heating buildings. This is what makes life tolerable on this island of ice near the Arctic Circle.

The Vikings, who originally discovered the island in the tenth century A.D. must have been intrigued by the energetic displays of hot water and steam gushing out of the ground, and indeed, the term *geyser* is from the Nordic word *geysir*, meaning gusher. The capital city of Reykjavik, which means "smoking bay," was named by the original Vikings because of its numerous whisps of steam rising from the surface of the water. Although Iceland

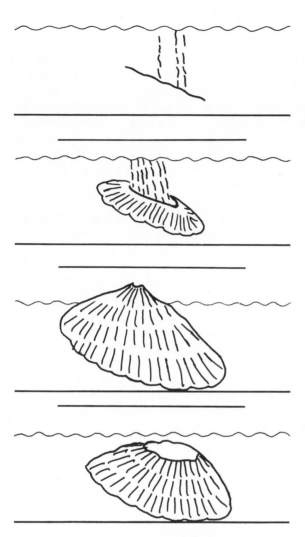

FIG. 5-1. Life cycle of an oceanic volcano.

is fortunate to have such a free supply of energy, it is not without its dangerous side effects, and the island has also been plagued with frequent volcanic eruptions.

The most destructive eruption in modern times occurred on the island of Heimaey near the fishing port of Vestmannaeyjar on January 23, 1973. It destroyed one-third of the town. Since then, there have been at least a dozen or more volcanic eruptions throughout the small country.

Iceland derives this mixed blessing from the mid-Atlantic spreading ridge system which gave the

island its existence. On other parts of the midocean ridge, volcanic activity is quite prevalent, and there are perhaps as many as 20 major eruptions a year. However, because these eruptions occur deep underwater their effects are not noticeable. Because Iceland is a surface expression of the Mid-Atlantic Ridge and straddles both sides of the rift, eruptions from volcanoes and geysers are more visible, but are not necessarily more frequent.

In 1783, when he was living in Paris, France, as the first diplomatic envoy of the United States of America, Benjamin Franklin noticed that there was a constant dry fog all over Europe and North America during the summer. This substantially reduced the temperature, and the winter of 1783-84 was one of the most severe winters on record. Franklin attributed the strange weather phenomenon to large quantities of ash from the Laki Volcano in Iceland, where 10,000 people and 200,000 livestock died from the eruption. The volcanic dust spread across the Northern Hemisphere during the summer of 1783 and blocked the light from the Sun, thereby shading the Earth.

Today, scientists have little doubt that the volcanic eruption in Iceland caused the bad weather that Benjamin Franklin spoke of. The eruption of Laki and other major volcanic eruptions are recorded in ice cores taken from the Greenland ice sheet. The cores can be accurately dated by counting the annual layers of ice. Those layers with a high acid concentration are indicative of large volcanic eruptions because volcanoes inject a large amount of sulfur compounds into the atmosphere, which rain out as sulfuric acid. From the years 1500 B.C. to A.D. 1500, the cores show five great European eruptions—three in the Mediterranean area, and two in Iceland. The ice cores plainly show another massive volcanic eruption that occurred in 1815, and this was the largest and produced the worst climatic effects of any volcano in modern history.

THE LARGEST AND DEADLIEST ERUPTION

Indonesia is known for its giant explosive volcanoes, which have produced more violent blasts in historic times than those of any other region (FIG. 5-3). They belong to a belt of 500 active volcanoes that surrounds the Pacific plate, called the "ring of fire" (FIG. 5-4). Three great volcanoes have made their presence known over the last two centuries: Mounts Tambora, Krakatau, and Agung. These and numerous smaller active Indonesian volcanoes are associated with a subduction zone called the *Java Trench* that runs from the tip of Burma to New Calendonia and is in the process of digesting the Indian-Australian plate.

Mount Tambora, occupying most of the Sanggar Peninsula on the island of Sumbawa, exploded on April 10 and 11, 1815. More than 88,000 people perished from direct and indirect effects on Sumbawa and neighboring Lombok Island. The eruptions began on April 5 with a series of thundering detonations that sounded like cannon fire and were heard as far away as the Molucca Islands off the tip of Malaya 900 miles away. Convinced that the island was being invaded by pirates, the Dutch sent troops into the area. On the following morning, ash began to fall on eastern Java, accompanied by less frequent detonations.

About 7:00 P.M. on April 10, the eruption intensified and attained its maximum convulsions. Three columns of flame rose up from the crater to very great heights, and soon the whole mountain appeared as though it were a mass of flowing liquid fires. Eight-inch-diameter pumice stones rained down on Sanggar, 20 miles to the east. Meanwhile, violent winds uprooted trees and other exposed objects as the volcano's massive eruption column collapsed under its own weight, forming a large caldera. Hot pyroclastic flows cascaded down the mountain and wiped out the village of Tambora, 12 miles to the south.

Very loud explosions were heard all night throughout Java, and concussions could be heard on the northwest end of Sumatra, 1600 miles away. The ash veil spread out over a vast area, extending as far as western Java and southern Celebes. In many areas within a 375-mile-radius of the volcano, the sky remained pitch dark for one or two days. Enormous rafts, some measuring over 1 mile across and several feet thick, were composed of pumice mixed with ashes and uprooted trees and floated in the sea to the west. A 3- to 12-foot tsunami, traveling at speeds up to 160 miles per hour, struck the shores of the Indonesian islands. The wave was probably

(Courtesy of USGS)

FIG. 5-2. Submarine eruption of Myojin-sho Volcano, Izu Islands, Japan.

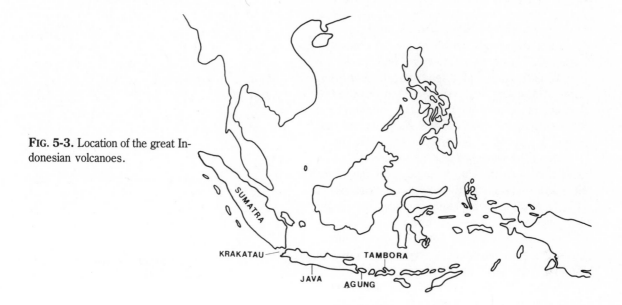

FIG. 5-3. Location of the great Indonesian volcanoes.

SUMATRA

KRAKATAU

JAVA

TAMBORA

AGUNG

FIG. 5-4. The ring of fire.

caused by the sudden entry of pyroclastic flows into the sea, rather than the subsidence of land caused by earthquakes associated with the eruption.

What was remarkable about Tambora was that it exceeded any other known eruption during the past 10,000 years, and it sent more dust into the upper atmosphere and obscured the sunlight more than any in the past 400 years. The eruption blew off the upper two-thirds of the mountain and cast some 25 cubic miles of debris into the atmosphere.

Its effects were not just felt locally, but had a large impact on the climate throughout the entire Northern Hemisphere. By the following summer of 1816, the ash had completely encircled the Earth, dropping temperatures as much as 7 degrees Fahrenheit in New England and 5 degrees Fahrenheit or more in Europe. The event went down in the annals as the "year without a summer."

In New England, spring was late, and when it finally arrived and crops began to grow, a killing frost in June took all but the hardiest of plants. When harvest was about to begin in the fall, a cold wave out of the north brought widespread killing frosts and finished off those crops that managed to survive the ordeals of the summer. The Europeans had it even worse as many parts of Europe were ravaged from the Napoleonic wars that ended in 1815. The scarcity of food brought on insurrections, riots, and eventually disease that killed over 100,000 people.

THE LOUDEST NOISE KNOWN TO MAN

The island of Krakatau (or Krakatoa) lies between Java and Sumatra in the Sundra Strait, a 15-mile-wide gateway from the Indian Ocean to the South China Sea and a key shipping route. Beginning on May 20, 1883, ships passing through the busy strait noticed the stirrings of the 2675-foot volcano that dominated the island and that had been quiet for the past 200 years. A great column of dust and smoke rose above Krakatau, and was carried by the wind to towns more than 100 miles away. The noise was like distant artillery fire and could be heard 70 miles away.

Throughout the summer, the volcano continued to rumble, and earth tremors were felt in many parts of Java and Sumatra. The main crater was blown

FIG. 5-5. Krakatau Island. Dashed line indicates caldera left after the explosion.

away and in its place were as many as ten subsidiary craters, all belching steam and ash that fell on villages 300 miles away. By Sunday afternoon, August 26, the violence of the eruption reached such proportions that the 36,000 inhabitants of Jakarta, the capital of Java 100 miles to the east, were overwhelmed by a rain of hot ash, and the booming sound was loud enough to rattle windows.

In the early morning of August 27, a series of four explosions ripped the island apart. The third explosion was by far the greatest. Apparently, the explosions were partially powered by the rapid expansion of steam, which was generated when seawater entered a breach in the magma chamber. Following the last convulsion, the majority of the island caved into the emptied magma chamber and created a large undersea caldera more than 1000 feet below sea level. The caldera looked somewhat like a broken bowl of water with jagged edges protruding above sea level (FIG. 5-5).

The explosions were equivalent to 3000 atomic bombs going off in a span of 5 hours. The sound from the explosions carried to Australia, Sri Lanka, and as far as Madagascar, 3000 miles away. The pressure wave was recorded on barographs all around the world, and it circled the globe at least three times, taking as long as 4 days to die down.

An estimated 36,000 people died in the immediate vicinity of the volcano. Most of the damage and loss of lives followed in the wake of giant tsunamis as high as 100 feet. For over 100 miles around Krakatau, the tsunamis washed over low-lying areas and flattened everything in their paths. One giant wave even carried a Dutch gunboat well over a mile inland. Villages and towns along the seashore were inundated with seawater, and inhabitants were carried out to sea to a watery grave. It was estimated that another 50,000 people died in this manner, bringing the total death toll close to 90,000 people.

The tsunamis registered on tidal gages as far as the English Channel, one-third the distance around the world. Two-thirds of Krakatau disappeared; much of it, about 7 cubic miles worth, went into the atmosphere. Rising some 50 miles in altitude, the dust circled the Earth and caused a 1 degree Fahrenheit temperature drop in the Northern Hemisphere. As a reminder of the volcano's destruction, for several years afterwards the dust produced some of the most magnificent sunsets the world has ever known.

CRACKS IN THE OCEAN FLOOR

Most of the volcanic activity that continually remakes the surface of the Earth takes place at the bottom of the ocean. Almost all of this activity is confined to the margins of lithologic plates. At convergent plate boundaries, where one plate is subducted under another, magma is formed by the melting of lighter constituents of the subducted plate. The upwelling magma can create island arcs, which are mostly in the Pacific. Some arcs include Indonesia, the Philippines, Japan, the Kuril Islands, and the Aleutians, the longest arc, which extends for more than 3000 miles. The island arcs have similar graceful curves, and each is associated with an ocean trench.

The curvature of the arcs might be caused by the slow rotation of the Pacific plate around a common pole. This would also set up stress fractures on the rim of the plate where it comes in contact with other plates. It might also account for why 70 percent of the world's active volcanoes, along with most earthquakes, exist on the edges of the Pacific plate.

On land, the upwelling magma can create volcanic mountain chains such as the Andes of South America and the Cascade Range of North America. Lavas associated with convergent plates, called *andesite lavas*, differ from the basalt lavas of the midocean ridges in that they contain more silica, calcium, sodium, and potassium and less iron and magnesium.

Along the midocean ridges, magma wells up from the mantle and spews out onto the ocean floor, where the spreading tectonic plates grow by the steady accretion of solidifying magma. The magma also flows out of isolated volcanic structures called *seamounts* (FIG. 5-6), which are strung out in chains across the interior of plates (FIG. 5-7). The mantle material extruded on the surface is basalt, which is rich in silicates of iron and magnesium.

The magma from which basalt is formed originated in a zone of partial melting in the Earth's upper mantle more than 60 miles below the surface. At this depth, the semimolten rock is less dense than the surrounding mantle material and rises slowly toward the surface in the form of giant blobs called diapirs. As the diapir ascends, the pressure decreases and more mantle material becomes molten. The rising diapir contributes to the formation of shallow magma reservoirs or feeder pipes that are the immediate source of volcanic activity.

The magma chambers closest to the surface are under the midocean ridges where the crust is only 6 miles thick or less. Not all of the magma is extruded onto the ocean floor; some solidifies within the conduits above the magma chamber and forms elongated structures known as dikes.

There are two types of volcanic eruptions associated with midocean ridges: fissure eruptions, and those that build typical conical volcanic structures. With the first type, the magma oozes onto the ocean floor in the form of lava through long fissures in the trough between ridge crests and along lateral faults. The faults usually occur at the boundary between tectonic plates, where the brittle crust is split apart by the separation of the plates. Magma welling up along the entire length of the fissure forms large lava pools (FIG. 5-8), similar to those of large, broad-shield volcanoes such as the Hawaiian volcano Mauna

Loa, the largest of its kind in the world (FIG. 5-9).

The two main types of lava formations in the midocean ridges are sheet flows and pillow, or tube, flows. Sheet flows are more prevalent in the active volcanic zone of fast-spreading ridge segments, such as those of the East Pacific Rise where in some places, the plates are separating at a rate of 6 inches per year. The sheet flows consist of flat slabs of basalt usually less than 8 inches thick.

Pillow lavas (FIG. 5-10) erupt as though basalt were squeezed out of a giant toothpaste tube. They are often found in slow-spreading ridges such as the Mid-Atlantic Ridge, where plates are separating at a rate of only about 1 inch per year and the lava is much more viscous. The surface of the pillows often has corrugations or small ridges, pointing in the direction of flow. The pillow lavas typically form small, elongated hills, pointing downslope.

When the upwelling magma is concentrated in comparatively narrow conduits that lead to the main feeder column, seamounts are formed. Some seamounts are associated with extended fissures along which magma welled up through a main conduit, piling successive lava flows on top of one another. The summit of a seamount sometimes has a crater, or depression, within which lava is extruded. If the crater exceeds 1 mile in diameter it is called a *caldera*, and its depth might vary from 150 to 1000 feet. Calderas are formed when the magma reservoir empties, creating a hollow chamber; without support, the top of the volcanic cone collapses, forming a wide depression. Feeder vents along the periphery of the

(Courtesy of NOAA)

FIG. 5-6. Divers exploring summit of the Cobb Seamount, located 270 miles southwest of Seattle.

FIG. 5-7. Linear chains of volcanic islands and seamounts in the Pacific.

FIG. 5-8. Rim of an undersea lava-lake collapse pit.

(Courtesy of USGS)

Cracks in the Ocean Floor **55**

caldera supply fresh lava that eventually fills the caldera and gives the volcano a flat-top appearance. Other undersea volcanoes do not have a collapsed caldera, but instead, the summit consists of a number of isolated volcanic peaks upwards of 1000 feet high.

Seamounts associated with midocean ridges that grow tall enough to break the surface of the ocean become volcanic islands. The volcanic islands associated with the Mid-Atlantic Ridge system include Iceland, the Azores, the Canary and Cape Verde islands off West Africa, Ascension Island, and Tristan da Cunha. The volcanic islands associated with the East Pacific Rise are the Galapagos Islands west of Ecuador.

Volcanoes formed on or near the midocean ridges can develop into isolated peaks as they move away from the axis of the ridge during seafloor spreading. The ocean floor thickens as it leaves the accreting plate boundary. This process can influence the height attained by the volcanoes as they move away from the spreading ridge axis because the thicker crust can support a greater mass on the ocean floor. However, a volcano formed at the ridge cannot increase its mass unless it continues to have a source of magma after it has left the vicinity of the ridge. Sometimes, a volcano formed on or near a midocean ridge develops into an island, but only if it is continually supplied with magma from an underlying magma chamber or feeder pipe. Most volca-

(Courtesy of USGS)

FIG. 5-9. Mauna Loa Volcano, Hawaii.

noes never make it to the surface of ocean and remain as isolated undersea volcanoes.

Because the crust under the Pacific Ocean is more volcanically active, it has a higher density of seamounts than the Atlantic or Indian oceans. The number of undersea volcanoes increases with increasing crustal age and thickness. The tallest seamounts, over 2.5 miles above the seafloor, are located in the western Pacific near the Philippine trench, where the crust is more than 100 million years old. The average density of Pacific seamounts is between five and ten volcanoes per 5000 square miles of ocean floor, a considerably greater occurrence of volcanoes than there is on the continents.

THE WORLD'S VOLCANIC HOT SPOTS

In addition to rift volcanoes, which obtain their source material from spreading ridges, and volcanoes associated with subduction zones, volcanoes also form in the interior of tectonic plates far away from centers of volcanic activity. These volcanoes develop where the upwelling of hot mantle material rises to the surface as the plate is moving over a fixed zone of partial melting in the upper mantle, called a *hot spot*. Throughout the world, there are some 40 active hot spots, many of which, like the one under Yellowstone National Park, exist on the continents. The upwelling currents also form broad domelike areas, or swells, in the ocean floor and under the continents. They average 700 miles across and account for about 10 percent of the earth's surface.

Often, the passage of a plate over a hot spot results in a trail of identifiable surface features whose linear trend reveals the direction in which the plate is moving. This trail produces volcanic structures that are aligned in a direction that is oblique to that of the adjacent midocean ridge system, rather than being roughly parallel to the spreading centers as are rift volcanoes. The hot-spot track might be a continuous volcanic ridge or a chain of volcanic islands and seamounts, rising high above the surrounding seafloor.

(Photo by F.H. Noffit, courtesy of USGS)

FIG. 5-10. Pillow lava on Knight Island, Alaska.

The most prominent example is the Hawaiian Islands, where the youngest and most volcanically active is Hawaii to the southeast, with progressively older islands having extinct volcanoes trailing off to the northwest. From there, coral atolls, such as Midway Island, and shoals were formed by successive layers of coral grown on the flattened tops of volcanoes that were worn down below sea level. Continuing where these islands end is an associated chain of undersea volcanoes called the *Emperor Seamounts*.

The Hawaiian Islands also lie parallel to two other island chains: those of the Austral Ridge and the Tuamotu Ridge. The islands and seamounts formed conveyer-belt fashion by the northwestward motion of the Pacific Plate. The plate did not always travel in this direction, and 40 million years ago, it followed a course with a more northerly heading. The course change, possibly as a result of a collision between the North American Plate with the Pacific Plate, shows up as a north-trending bend in the hot-spot tracks. Thus, the existence of these islands and many more is used as further evidence for the theory of plate tectonics.

The hot spots derive their source material from deep inside the mantle. The distinctive composition of the hot-spot lavas indicate that their source is isolated from the general circulation of the mantle. As plumes of this mantle material flow upward into the asthenosphere, a region where rocks are plastic, the part rich in volatiles rises to the surface to feed the hot-spot volcanoes. The plumes come in a range of sizes that might be indicative of the depth of their source material. The plumes might not necessarily be continuous flows of mantle material, but instead consist of molten rock, rising in blobs or diapirs. Sometimes a hot spot will fade away entirely, and a new one be formed in its place. The typical life span of a plume is on the order of 100 million years.

The position of a hot spot can change slightly, and as a result, the tracks on the surface are not all as linear as the Hawaiian chain. Compared with the plates, however, the mantle plumes are relatively stationary. Because the motion of the hot spots is insignificant, they provide a reference point for determining the direction and rate of travel of a plate.

If a midocean ridge passes over a hot spot, the plume directly under the spreading center augments the flow of molten rock, welling up from the asthenosphere to form new crust. The crust over the hot spot is therefore thicker than it is along the rest of the ridge, resulting in a plateau rising above the surrounding sea floor. The most striking example of this is Iceland, which straddles the Mid-Atlantic Ridge and was born a mere 15 million years ago. There, the upwelling is so intense and the crust is so thick that the plateau rises above sea level. In both directions along the ridge, the abnormally elevated topography extends out to a distance of about 900 miles, only 350 miles of which lie above sea level. South of Iceland, the broad plateau tapers off to form the typical Mid-Atlantic Ridge. The powerful upwelling currents also produce glacier-covered volcanic peaks up to 1 mile high. In a geologically brief period, the Mid-Atlantic Ridge will move away from the hot spot, carrying Iceland along with it. Devoid of a source of magma, the previously active volcanoes will cease their eruptions, and Iceland will become just another cold, ice-covered rock.

The World of Ice

ROUGHLY 10 percent of the Earth's surface is covered by ice (FIG. 6-1), and glaciers are found on every continent. About 3 percent of all the water in the world is freshwater, and about three-quarters of that is locked up in glacial ice. Excluding the polar ice caps, the remaining glaciers on the continents hold as much fresh water as do all the world's rivers and lakes.

Since the end of the Cretaceous period 65 million years ago, when the dinosaurs became extinct, the world for the first time formed two permanent polar ice caps. Throughout the Earth's long history, just having a single polar ice cap was a rare event. Most of the time, the Earth was warmer than it is today and essentially ice free.

During the most recent ice epoch, spanning the last 2 million years, numerous ice ages have come and gone almost on a regular basis. The last ice age, when one-third of the earth's land surface was covered with ice, peaked around 18,000 years ago and ended about 10,000 years ago. Today, we live in the lull between ice ages, called an *interglacial* age. The present interglacial age is almost over. It would be interesting to look several thousand years into the future and see just how well modern man copes with the next ice age.

THE GIANT ICE BALL

Over the past 2 billion years, there have been four major ice epochs. In geology, an *epoch* is a span of time that is less than a period but greater than an age, and can cover several millions of years. Presently, we live in an ice epoch that so far has lasted 2 million years. Within this time were ice ages, having roughly 100,000-year cycles. Each ice age persisted for about 90,000 years and was followed by a short, warm interglacial age that lasted about 10,000 years. The fact that the last ice age ended 8000 to 10,000 years ago indicates that the present interglacial probably has less than 2000 years to go before the beginning of a new ice age.

The ice age might come on very slowly and over the short run be hardly noticeable. It might take tens of thousands of years before the ice sheets reach their maximum extent. They would then be followed by a rapid recession until most of the Earth again would be free of ice. During the last ice age, it took over 70,000 years for the ice sheet to reach its peak and less than 10,000 years for it to retreat to its present position.

The ice epoch we are presently living in is relatively mild compared to the great ice epochs of the distant past. Two billion years ago, over half the globe was encased in ice. This was a transition period when the composition of the atmosphere was changing from carbon dioxide to oxygen. This was also about the time when plate tectonics began to operate extensively, scrubbing out carbon dioxide by locking it up in carbonaceous sediments and burying the rocks deep within the interior of the Earth.

A second major ice epoch occurred about 1 billion years ago, and this one was perhaps the greatest glaciation the planet has ever endured (FIG. 6-2).

The climate was so cold that ice sheets and permafrost (soil frozen throughout the year) lay near the equatorial latitudes. No plants grew on the barren land surface, and no animal life lived in the seas at that time.

Thick sequences of Precambrian *tillites*—a mixture of boulders, pebbles, and clay consolidated into solid rock—are known to exist on every continent. In the Lake Superior region, tillites are 600 feet thick in places and range from east to west for a distance of 1000 miles. In northern Utah, tillites mount up to a total thickness of 12,000 feet and give every impression that there were a series of ice ages following one another in close succession. Similar tillites have been found among the Precambrian rocks of Norway, Greenland, China, India, southwest Africa, and Australia.

(Courtesy of U.S. Navy)

FIG. 6-1. U.S. Navy submarine in pancake ice, Arctic Ocean off Alaska.

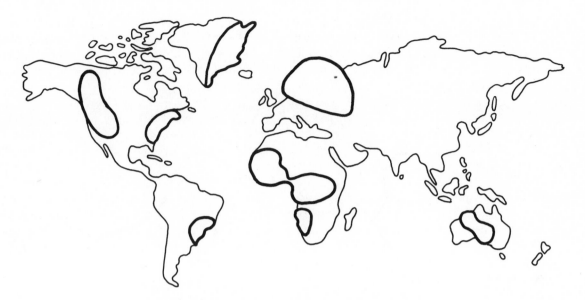

FIG. 6-2. Worldwide extent of late Precambrian glaciation.

Following the retreat of the ice, life began to proliferate in the ocean on a grand scale. As the oceans began to warm, life processes speeded up and organisms evolved into more complex forms. By the end of the Proterozoic eon for the first time, large numbers of fossils of multicellular animal life were preserved, indicating a sharp demarcation between the Proterozoic and the Phanerozoic eons.

The third great ice epoch took place toward the end of the Paleozoic era, about 240 million years ago. All the southern continents were assembled into one supercontinent. When Gondwanaland separated from its northern counterpart, Laurasia, and passed into the south polar regions, glacial centers expanded in all directions (FIG. 6-3). Ice sheets covered large portions of east-central South America, Antarctica, South Africa, India, and Australia. In Australia, scientists have found marine sediments interbedded with glacial deposits and tillites separated by seams of coal, indicating that the ice epoch was interspersed with warm interglacial spells. In South Africa, late Paleozoic tillites several thousand feet thick cover several thousands of square miles.

Early into the ice age, the maximum glacial effects were in South America and South Africa. Later, the chief glacial centers moved to Australia and Antarctica, providing good evidence that these continents were at one time locked together and wandered around the South Pole.

OF ICE AND MEN

Our present ice epoch takes place within a period known as the Quaternary, which consists of the Pleistocene epoch (the age of recent life) and the Holocene epoch (concurrent with the age of modern man, the last 10,000 years after the last ice age). The Pleistocene epoch also has become synonymous with the most recent ice ages.

What is outstanding about this particular ice epoch is that our ancestors appeared on the scene just about when the ice ages began some 2 million years ago, and ice ages span the whole of man's existence. This is not to say that man coexisted with the ice, for the cradle of civilization was far removed from the glaciated areas, and man evolved in a region of Africa that was actually quite warm. Perhaps during the interglacial periods, man wandered farther north into Europe and Asia, only to be chased out again by the next advancing ice sheet.

It has only been within the latest interglacial period that man has made permanent residences in

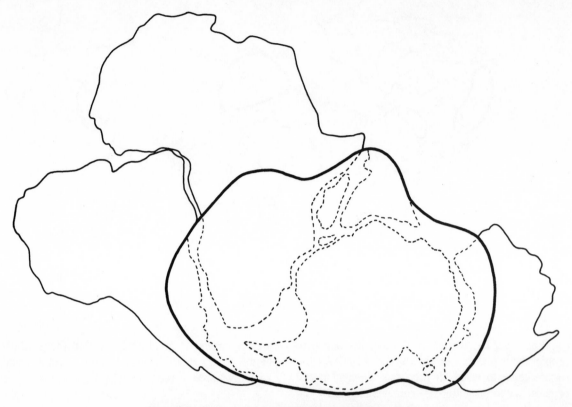

Fig. 6-3. Extent of late Paleozoic glaciation in Gonwanaland.

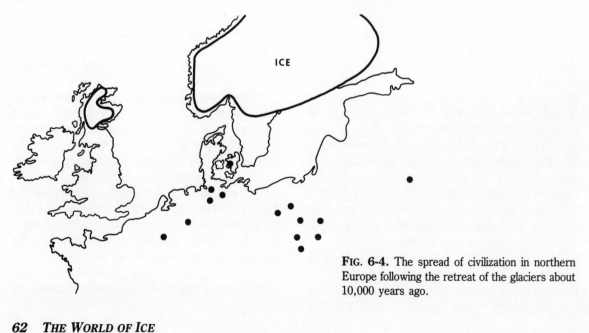

ICE

Fig. 6-4. The spread of civilization in northern Europe following the retreat of the glaciers about 10,000 years ago.

FIG. 6-5. Worldwide extent of Pleistocene glaciation.

the north (FIG. 6-4). It was warm enough for the Vikings to settle and thrive in Greenland (that is how it got its name) beginning about A.D. 950. Then, in a period known as the "Little Ice Age," from about 1400 to 1850, creeping glaciers froze out the Greenland Vikings and chased people out of the northlands of Europe.

At the height of the last ice age (FIG. 6-5), ice was piled as high as 10,000 feet over Canada, Greenland, and northern Europe. In North America, an ice sheet covering 5 million square miles spread out from Hudson Bay, reaching north to the shores of the Arctic Ocean and south to bury all of eastern Canada, New England, and much of the rest of the northern half of the midwestern United States. A smaller ice sheet, originating in the Canadian Rockies, engulfed western Canada, parts of Alaska, and small portions of the northwestern United States. Ice buried the mountains of Wyoming, Colorado, and California, and rivers of solid ice linked them with mountains in Mexico.

In Europe, the ice sheet radiated from northern Scandinavia and covered most of Great Britain and large parts of northern Germany, Poland, and the Soviet Union. A smaller ice sheet, centered in the Swiss Alps, covered parts of Austria, Italy, France, and southern Germany. In Asia, the ice sheets occupied the Himalayas and parts of Siberia.

In the Southern Hemisphere, only Antarctica had a major ice sheet, which had nowhere to go except into the sea. Small ice sheets expanded in Australia, New Zealand, and the Andes of South America. Throughout the world, alpine glaciers (FIG. 6-6) existed on mountains that are presently ice free.

When the ice age was well entrenched, the average global surface temperature was about 10 degrees Fahrenheit colder than it is today. About 5 percent of the world's water was locked up in ice, dropping the sea level 300 feet lower than it is today. The ice reached its greatest extension 18,000 years ago and began a rapid melting between 16,000 and 13,000 years ago, during which time one-third of the ice melted. The melting of the ice sheets appeared to pause between 13,000 and 10,000 years ago. Then a second episode of melting led to the present volume of ice by about 6000 years ago.

The two stages of deglaciation might have come

about from the flooding of large numbers of icebergs, which calved off the ice shelves of the Arctic during the first melting stage and chilled the water of the North Atlantic, ultimately cooling the Northern Hemisphere (FIG. 6-7). At the same time, the rising sea level also led to the creation of high-speed ice streams, flowing from the central ice sheets into the sea. The ice sheets rapidly thinned without a marked change in their extent and collapsed from within. This might explain why the ice age ended as abruptly as it did, even though the temperature was not significantly warmer than it is today.

IS THE SUN FADING?

The measurement of the solar brightness, or *irradiance*, taken by the Solar Maximum Mission (Solar Max) satellite launched in 1980 indicated a to-tal decrease in solar irradiance between the beginning of 1980 and early 1985 of nearly 0.1 percent. That is quite a drop for a property that astronomers thought to be so stable they called it the ''solar constant.''

One explanation for the apparent decrease in solar output is that the number of sunspots has been decreasing from their peak in 1980. The trend suggests a possible association with the 22-year solar cycle, during which time sunspot numbers peak every 11 years and change their magnetic polarity every 22 years. In the interval between maximum and minimum sunspot numbers, the overall reduction of irradiance could be 0.2 percent. Keep in mind that a decrease as small as 0.1 percent over a decade or more could have a significant influence on Earth's climate.

(Photo by T.L. Powe, courtesy of USGS)

FIG. 6-6. Maclaren Glacier, Alaska.

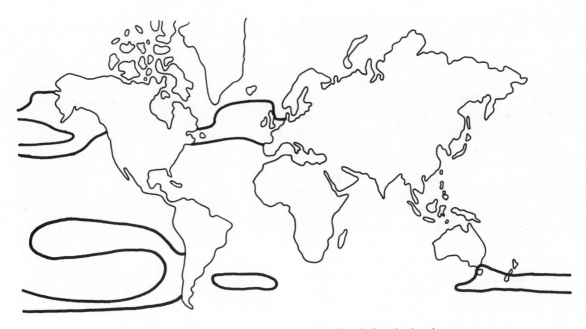

FIG. 6-7. Areas of August sea-surface cooling during the last ice age.

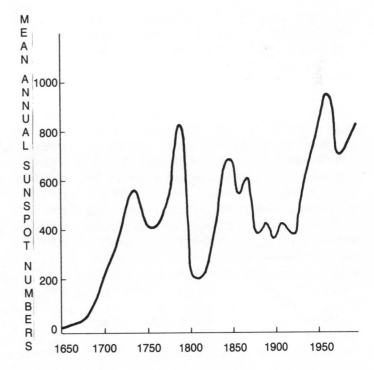

FIG. 6-8. Sunspot activity through time.

One possible irradiance-climate connection goes back to the sixteenth and seventeenth centuries, when Europe and other regions were in the grip of a Little Ice Age. During this time, sunspots were rare (FIG. 6-8), an event known as the *Maunder Minimum*, and the decrease in irradiance might have been as much as 1 percent, which would have caused an estimated 2 degrees Fahrenheit cooling.

Although sunspots do mar the surface of the Sun, they do not block solar radiation in the direction of the Earth, as it was once thought. One interesting solar phenomenon that could be involved is an apparent expansion of the Sun, and there could be a link between the Sun's size and its irradiance.

Evidence for a variable Sun is found in Precambrian lake-bed sediments, called *varves*, located in South Australia, north of Adelaide. What makes these rocks so remarkable is that they incorporate a complex pattern of laminations whose thicknesses might record cyclic variations in mean annual temperature over many thousands of years. The patterns might have been caused by cyclic variations in the activity of the Sun that are strikingly similar to the cyclic activity the Sun exhibits today, almost 700 million years later. The sediments were laid down during the late Precambrian ice epoch when periodic influxes of glacial meltwaters turbid with sediments flowed into a large lake. The cyclic variations of meltwater left successive silty laminations deposited on the lake bed.

What was astonishing about these alternating bands of red siltstones and fine sandstones is that they mimic both the 11-year sunspot cycle and the 22-year solar cycle. The similarities argue for a direct connection between varve thickness and solar activity. An increase in solar activity caused a corresponding increase in climatic temperature, which in turn entailed a greater annual discharge of meltwaters and the deposition of thicker varves on the lake bed. The data also implies that the Sun's activity has not changed significantly over the past 700 million years, indicating an apparent stability of the Sun's behavior.

Trees are also reliable climatic indicators, and the wider their tree rings are, the more hospitable the climate was for the year in question. Individual

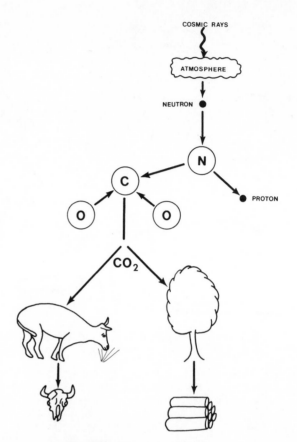

FIG. 6-9. The carbon-14 cycle.

rings are also analyzed for their carbon-14 content. Ordinary carbon-12 along with radioactive carbon-14 are incorporated in the tree's tissues during its growth. Carbon-14 has a half-life (the time it takes for half the atoms to decay into a stable element) of 5730 years. It is generated in the atmosphere by the bombardment of nitrogen-14 with cosmic rays (FIG. 6-9). By measuring the carbon-14 content of the rings of ancient, well-preserved trees, scientists are able to reconstruct the history of carbon-14 in the atmosphere, going back more than 7000 years.

A drought index for the western United States going back to the year 1600 also has been obtained by analyzing the width of tree rings of the bristlecone pine, one of the longest living plants on Earth. The tree rings showed a drought period roughly every 22 years that matches the 22-year solar cycle.

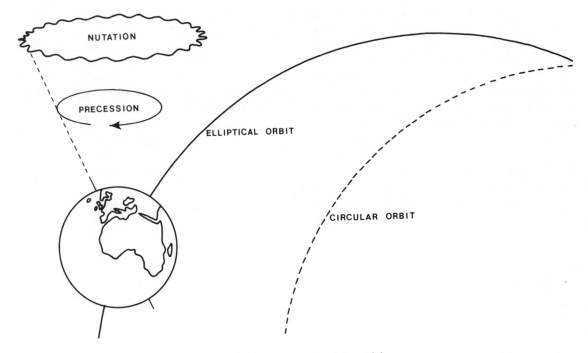

FIG. 6-10. The Milankovich model.

What was even more dramatic was that those trees living in the period between 1645 and 1715 showed in their rings an account of very unusual climatic activity that corresponds precisely with the Maunder Minimum of sunspot activity.

A COSMIC SUN DANCE

Over the past million years, there have been at least nine ice ages, each coming and going as though they were as cyclical as the seasons. Although many theories have been proposed to explain the ice ages, only one seems to have a convincing argument. This theory purports that ice ages were caused by small changes in the tilt of the Earth's axis and in the geometry of the Earth's orbit around the Sun. It was developed by the Yugoslav astronomer Milutin Milankovitch during the early part of this century (FIG. 6-10).

Milankovitch painstakingly calculated the changes of incoming solar radiation, latitude by latitude, and found there were three orbital cycles that coincided with the 100,000-, 41,000-, and 22,000-year ice age cycles. They are, in order, the shape of the orbit, the precession of the equinoxes, and the tilt of the axis. The theory asserts that orbital variations changed the climate by altering the amount of solar energy the Earth receives at different latitudes and seasons. Therefore, cool summers and not cold winters are enough to bring on an ice age.

Unfortunately, when Milankovitch proposed his theory in 1941, there was no means of testing it because there was no accurate method of dating the ice ages, so it was essentially ignored until the late 1960s. Recently, however, new types of geologic evidence have been brought to the forefront, and a close correspondence has been demonstrated between the history of the ice ages and the orbital variations. It has also been satisfactorily explained how small variations in solar *insolation*—that is, the radiation received from the Sun—can cause large changes in the climate.

One method of pinning down the dates of the ice ages is by measuring the amount of global ice at any one time. This is accomplished by analyzing the ratios of two isotopes of oxygen in ocean sediments deposited during that time. Most of the oxygen in the ocean is oxygen-16, but there is also a tiny amount of heavy oxygen-18. The heavier molecules of oxygen tend to be left behind when seawater evaporates. As a result, the precipitation on land has less oxygen-18 than does seawater. When an ice age begins and the continental glaciers grow, water is removed from the ocean and stored in the ice. The seawater left behind thereby becomes enriched in the heavier oxygen-18. Meanwhile, marine organisms construct their shells, using the available oxygen atoms in the seawater, which reflect the water's isotopic composition. When the organisms die, their shells fall to the ocean floor, where they accumulate to form carbonaceous sediments. A higher ratio of oxygen-18 to oxygen-16 in a sediment sample would therefore indicate more land ice at the time when the sediment was laid down. The comparison of oxygen isotope levels in ice cores from the Greenland and Antarctic ice sheets are another means of measuring paleotemperatures.

Another method of dating the ice ages is by dating coral terraces, which resemble a staircase of coral growth running up an island or a coral reef. During the ice ages, a lot of water is locked up in the ice sheets on land and the sea level lowers appreciably. Since coral only grows at sea level, fluctuations in sea level can be accurately determined by age-dating the coral.

A plot of the age dates indicates that the global ice volume has fluctuated repeatedly every 100,000 years for the past several hundred thousand years. The curves of the cycles also show that the ice took much longer to build up than it did to disappear. In addition, there were smaller fluctuations superimposed on the principal 100,000-year cycle. The most obvious conclusion that can be drawn from these facts is that the climate is influenced by cyclical variations in the amount of sunlight the Earth receives. Since it has been shown that the Sun, whose small variations in output also run in cycles, has been fairly steady over the past billion years, the Milankovitch theory proposes a better reason for the variation of sunlight: the variation of the distance between the Earth and the Sun and the orientation of the Earth with respect to the Sun.

Near the height of the last ice age, the North Pole did not point toward Polaris, the North Star, as it does today. Instead, it pointed to Vega, the brightest star in the constellation Lyra. Therefore, the Earth's axis of rotation was 47 degrees in the opposite direction it is now and the seasons were reversed, with the northern winter occurring between July and October. Those constellations presently seen only in the Southern Hemisphere were then seen in the Northern Hemisphere.

The precession of Earth's axis is like the wobble of a toy top, tracing a complete circle in the heavens every 26,000 years. Together with the eccentricity of the Earth's orbit, which is the variation in the time of perihelion, or closest approach to the Sun, the cycle becomes approximately 22,000 years, somewhat shorter than the period of precession itself.

The tilt of the Earth's axis, which is responsible for the seasons, was not always 23.5 degrees, but varied between 22 and 24.5 degrees. The greater the tilt angle, the greater is the temperature difference between seasons because the axial tilt influences the angle at which sunlight impinges upon the Earth at various latitudes. A lesser angle is especially noticeable in the higher latitudes where ice forms easily. A complete tilt cycle from maximum to minimum and back to maximum tilt angle is called *nutation* (nodding up and down), and takes about 41,000 years. For the past 10,000 years, the degree of tilt has been getting progressively lesser, which should produce cooler summers and milder winters.

The Earth's orbit around the Sun ranges from nearly circular to highly elliptical. When orbiting the Sun in a circle, the Earth maintains a constant distance of 93 million miles from the Sun during all seasons, and the amount of heat received from the Sun is the same throughout the year. During an elliptical orbit, the Earth is millions of miles farther away from the Sun in one season and millions of miles closer to the Sun in the opposite season. Should the Northern Hemisphere be in summer when it is the farthest from the Sun, the lessened amount of sunlight might fail to melt the snow of the previous winter and more snow would be piled on top of it during the following winter.

Currently, the Earth is closest to the Sun on January 3 and farthest from the Sun on July 4, which means that the sunshine is 7 percent weaker during the northern summer than during the southern summer. A complete orbital cycle from near circular to elliptical and back again takes roughly 100,000 years, and this has perhaps the greatest effect on the waxing and waning of the great ice ages.

ICE ON THE BOTTOM OF THE WORLD

The continent of ice was discovered scarcely more than two centuries ago, and then it was stumbled upon by accident, even though its existence was predicted by Greek scholars over 2000 years earlier. The British navigator James Cook discovered *terra incognita*, or unknown land, in 1774, although heavy pack ice forced him to turn back before he actually set eyes on the frozen continent. By 1820, sealers were hunting seals, which were prized for their oil as well as their pelts, in the frigid waters around Antarctica.

The United States, Great Britain, France, and Russia sent out exploratory expeditions that made the first official sightings of Antarctica. One of these expeditions was commanded by the Scottish explorer Sir James Clark Ross, who in 1839 attempted to find the South Magnetic Pole. He drove his ships through 100 miles of pack ice on the Pacific side of the continent until finally emerging into open water, known today as the Ross Sea. Finding his way blocked by an immense wall of ice 200 feet high and 250 miles long, Ross had to give up his quest to the South Magnetic Pole, which lay some 300 miles inland from his position.

In 1902, the British explorer Commander Robert Scott attempted and failed to reach the geographical South Pole from McMurdo Sound (FIG. 6-11), turning back after covering only about one-third the distance to the pole. In 1909, one of Scott's former team members, the British explorer Ernest Shackleton, came within 112 miles of the pole, but had to turn back because of low supplies and foul weather. As a consolation, one of Shackleton's teams did find the South Magnetic Pole, a less glamorous prize, but an important scientific achievement nonetheless.

In 1911, Scott made a second attempt to the Pole, but this time he had competition from the Norwegian explorer Roald Amundsen. The Norwegians were the first to reach the South Pole on December 15, completing the 1600-mile round trip in less than 100 days. Scott reached the Pole a month later. On his return, he and two of his companions were caught in a blizzard and froze to death just 13 miles from their supply depot.

Almost 90 percent of all the ice in the world lies atop Antarctica. With an area of 5.5 million square miles, this desolate world of ice is larger than the United States, Mexico, and Central America combined. The ice rises more than 2 miles in places, and has an average thickness of 1.3 miles. There is about 7 million cubic miles of ice. To place this number into some sort of perspective, think of a block of ice roughly 200 miles long, 200 miles wide, and 200 miles high.

Despite all this ice, Antarctica is literally a desert, with an annual snowfall of about 2 feet, which translates into about 3 inches of rain. The snow accumulates because there is virtually no melting from year to year. The mean monthly temperature at the South Pole in summer is −28 degrees Fahrenheit; in winter it is −80 degrees Fahrenheit. In some places, the temperature has been known to drop to −127 degrees Fahrenheit. Barren mountain peaks soar 17,000 feet above the ice sheet, and winds shriek off the ice-laden mountains and tall ice plateaus at velocities upwards of 200 miles per hour—stronger than those of hurricanes and typhoons that plague other continents.

Antarctica discharges over 1 trillion tons of ice into the surrounding seas annually, and the ice flowing into the ocean calves off to make icebergs (FIG. 6-12). Behind the Transantarctic Range, a wall of mountains that divides the continent between East Antarctica and smaller West Antarctica, rivers of ice slowly flow outward and down to the sea on all sides. The ice escapes through mountain valleys to the ice-submerged archipelago of West Antarctica, and to

FIG. 6-11. U.S. Coast Guard icebreaker escorts oil tanker into McMurdo Sound, Antarctica.

the great ice shelves of the Ross and Weddell seas.

The ice in East Antarctica is firmly anchored on land, but the ice in West Antarctica rests below water on an ocean shelf and is fringed by floating ice that is pinned in by small, isolated islands buried under the ice. Normally, the flow of ice into the Antarctic Ocean compensates for the accumulation of snow in the interior so that the ice sheet is perfectly balanced and does not significantly change its volume with time. As the ice sheet grows, it tends to cool the surrounding atmosphere, thereby reducing the amount of snowfall. However, the increased load of ice also depresses the continent, which lowers the ice-sheet surface and raises its temperature.

Thus the ice continually readjusts itself to minor climatic changes.

Antarctica was not always shrouded in ice. Eighty million years ago, it was still attached to Australia and South America. The water surrounding Antarctica was reasonably warm and there were no ice sheets. Evidence for a warm climate that supported lush vegetation and even forests exists in coal seams that run through the Transantarctic Mountains. These are some of the most extensive coal beds on Earth.

As Antarctica pulled away from South America and then Australia, its increasing isolation from tropical heat sources allowed the cold circum-Antarctic

(Courtesy of U.S. Navy)

FIG. 6-12. U.S. Navy ships move large iceberg near McMurdo Station, Antarctica.

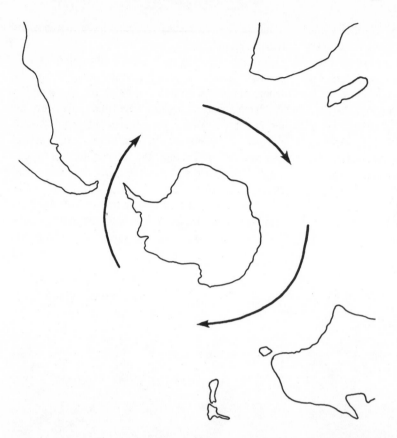

FIG. 6-13. The circum-Antarctic current.

current to encircle the continent (FIG. 6-13). The current presented a barrier to warm equatorial waters and a first, and probably the greatest, ice cap formed about 30 million years ago. Sometime during the next 15 million years, most of the ice sheet melted, possibly because of a warmer climate. Then about 15 million years ago, a new ice cap formed in its place as the climate became colder and the ocean-bottom temperature approached the freezing point of water. It is uncertain how many times the ice sheet has come and gone, but about 4 million years ago forests grew on the flanks of the Transantarctic Mountains, as indicated by discoveries of nonfossilized wood and marine fossils. In the relatively warm climate, great open seaways might have reached deep into the interior of the continent, and the central ice mass might have retreated to much smaller ice sheets and high alpine glaciers.

Most of the ice on Antarctica today accumulated during the last ice epoch. The ice is not uniform,

but instead is pervaded by internal layers. Large flat areas beneath the ice are thought to be subglacial lakes, kept from freezing by the interior heat of the Earth. The temperature a mile below the surface of the ice can be 25 degrees warmer than the temperature of the ice at the top. Add to that the high pressures that occur at such depths, and liquid water can exist at several degrees colder than its normal freezing point.

The pools of liquid water tend to lubricate the ice streams, which flow down the mountain valleys into the sea. This lubrication allows the ice streams, which might be several miles broad, to glide smoothly along the valley floors. The banks of the ice streams are flanked by deep crevasses where they make contact with the walls of the valley. Crevasses also run parallel to each other down the entire length of the ice streams. Normally, when placed under great stress, ice will shatter, but because of its great size, a glacier acts like a viscous solid and can flow. Most of the time the glaciers creep across the landscape at a snail's pace, gaining perhaps one-half mile per year. At other times, glaciers have been known to gallop, covering ground at an astonishing 100 feet per day.

One specter of the Antarctic ice that has some scientists worried is the possibility of an ice surge. A total collapse of the West Antarctic ice sheet could come about as the result of a warmer climate generated by an increase in atmospheric carbon dioxide. The rapidly melting, unstable ice sheet could come loose from its moorings and crash into the ocean. With the ice shelves gone, the rise in sea level worldwide would be upwards of 20 feet and would inundate coastal areas.

Even a slow melting of both polar ice caps could raise the level of the oceans upwards of 12 feet by the year 2100, drown much of the world's coastal plains and flood coastal cities. A rise in sea level also could lift West Antarctic ice shelves off the seafloor and set them adrift into equatorial waters, where they would rapidly melt and contribute to the calamity.

Rivers in the Sea

IN the days of old when ships depended on wind and sail to get around, a quite remarkable discovery was made by Benjamin Franklin when he was working for the London post office prior to the American Revolutionary War. British mail packets sailing to New England took 2 weeks longer to make the journey than did American merchant ships. American ships were no better than British ships, so the American sailors must have discovered a faster route—and indeed they had. American whalers first noticed that whales kept to the edges of what appeared to be an invisible stream in the ocean and did not attempt to cross it or swim against its current. Meanwhile, British captains, unaware of this stream, sailed in the middle of it. Sometimes if the winds were weak, the ships were actually carried backwards.

The current traveled 13,000 miles clockwise around the North Atlantic basin at a speed of about 3 miles per hour. In 1769, Franklin had the current mapped, thinking it would be a valuable aid to shipping. Considering the crude methods of chart making in his days, Franklin's map of the so-named Gulf Stream was unusually accurate. However, another century passed before any serious investigations of the current were ever conducted.

THE UNDERSEA CIRCULATORY SYSTEM

Shallow ocean currents (FIG. 7-1) are driven by the winds, whose momentum is imparted to the ocean's surface. The currents do not flow in the direction of the wind, but are deflected by the Coriolis effect to the right of the wind direction, or to the northwest, in the Northern Hemisphere and to the left of the wind direction, or to the southwest, in the Southern Hemisphere (FIG. 7-2). The currents pick up warm water from the tropics, distribute it to the higher latitudes, and return with cold water. This process moderates the temperatures of coastal regions and makes Japan and the British Isles warmer than they would otherwise be at their latitudes.

FIG. **7-1.** The ocean currents.

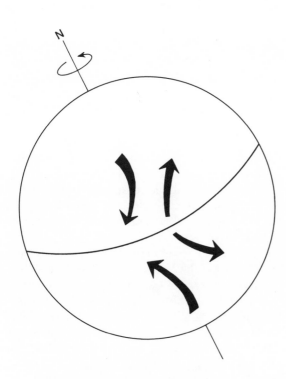

FIG. **7-2.** The Coriolis effect.

The Gulf Stream is also unstable, like its counterpart in the upper atmosphere, the jet stream. It snakes its way around the North Atlantic basin, surrounded by eddies or rings of swirling warm and cold water. These eddies are like underwater tornadoes; only some of these are enormous, as much as 100 miles or more across and reach down to depths as far as 3 miles. The eddies are probably pinched off sections of the main ocean currents. Their rotation is *anticyclonic*, or clockwise, in the Northern Hemisphere and *cyclonic*, or counterclockwise, in the Southern Hemisphere. Most eddies are less than 50 miles across, and some, like those found in the Arctic Ocean off Alaska, are only 10 miles across.

Like giant egg beaters, these small eddies play an important role in mixing the oceans. Marine life caught in these eddies are often transported to more hostile environments where they can only survive for as long as the food supply or the eddies hold out, which can take from several months to a year or more.

Cold water is denser than warm water, and the sinking of cold water near the poles generates strong, deep, steady currents that flow toward the equator. Associated with these currents are eddies on the western side of ocean basins that can be over

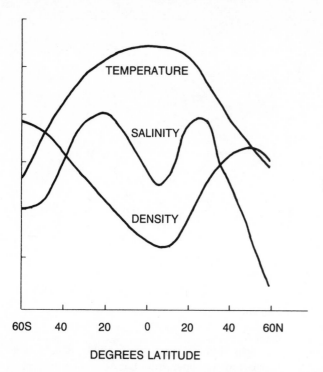

FIG. 7-3. Properties of the ocean by latitude.

TEMPERATURE

SALINITY

DENSITY

| 60S | 40 | 20 | 0 | 20 | 40 | 60N |

DEGREES LATITUDE

100 times stronger than the main current. The surface water in the polar regions is denser than in other parts of the world because it is both colder and saltier at the poles (FIG. 7-3). The saltiness comes from the evaporation of poleward-flowing water and the exclusion of salt from ice as it freezes. As the water increases in density, it sinks. It spreads out upon hitting the ocean floor, and heads toward the equator. Because of the Coriolis effect acting on the Earth's eastward spin, the currents are deflected westward.

The path taken by the circulating water is also controlled by the distribution of landmasses and by the topography of the ocean bottom, including midocean ridges and ocean canyons. Deep, cold currents, flowing from Antarctica toward the equator, are forced to the left and are pressed against the western side of the Atlantic, Pacific, and Indian ocean basins. This squeezing against the continents causes the currents to pick up speed, similar to the way a stream flows faster when its channel narrows.

The Indian Ocean is unique in that it is not connected to the north polar region and therefore has only one source of cold bottom water. The Atlantic and Pacific are both joined to the Indian and the Arctic oceans; however, the flow of deep cold water into the Pacific from the Arctic Ocean is diminished because the North Pacific is effectively blocked off from the Arctic Ocean by the shallow, narrow Bering Strait, which separates Alaska from Asia. In addition, the water of the Pacific is less salty than Atlantic water because of a smaller contribution of salt from rivers; therefore, it freezes more readily in the arctic regions. The unfrozen near-surface water is not sufficiently enriched in salt and not dense enough to sink to the bottom.

The surface water of the North Atlantic is saltier than that of the North Pacific. The Atlantic has two major sources of salty water. One is from the Gulf of Mexico, which is carried north by the Gulf Stream, and the other is the deep flow from the Mediterranean Sea. The surface water of the North Atlantic moves northward and enters the Norwegian Sea, where it is cooled below 32 degrees Fahrenheit, the freezing point of fresh water, but does not freeze because of its salt content. The cold, dense water sinks, and upon reaching the bottom, it reverses direction and flows back into the Atlantic through a series of narrow, deep troughs in the submarine ridges that connect Greenland, Iceland, and Scotland. This large volume of deep water moves south and is forced to the right against the continental margin of eastern North America, forming the so-called *western boundary undercurrent*. This current, recognized only about a couple of decades ago, transports 100 million cubic feet of water per second along the East Coast of the United States.

The deep ocean currents serve to distribute the cold waters from the poles to the tropics. If this were not so, the heat from the Earth's interior would make the ocean bottom warmer than the surface. This process could cause an entire ocean to overturn, bringing deadly amounts of carbon dioxide to the surface with disastrous consequences to sea life, and to the rest of life on Earth.

The sinking of water in the polar regions must be evenly matched with rising water in other parts of the world. The cold waters that sink in the polar seas rise in upwelling zones in the tropics, creating an efficient heat-transport system. The tropical seas are heated by solar radiation from above and cooled by upwelling water from below, giving rise to an equator-to-pole transport of heat from the oceans to the atmosphere. This involves the entire ocean waters in a gigantic heat engine that transports a great deal of heat. The deep-ocean currents move very slowly, completing the journey from equator to pole and back again in upwards of 1000 years. By comparison, surface currents can complete a circuit around an ocean basin in less than a decade.

Upwelling currents driven by offshore and onshore winds (FIG. 7-4) are also important for transporting bottom nutrients such as nitrates, phosphates, and oxygen to the surface. Microscopic plant life called phytoplankton thrive near the surface of the ocean where sunlight can penetrate for photosynthesis. These tiny organisms are at the very bottom of the food chain. As sea life flourishes, it depletes the water of its important nutrients which limits further growth of marine plant and animal populations. Fortunately, scattered around the world are numerous upwelling zones of colder, nutrient-

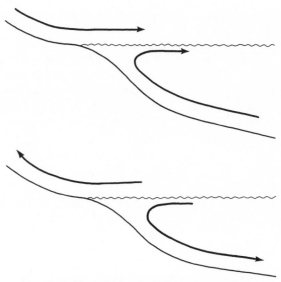

FIG. 7-4. Upwelling and sinking ocean currents.

rich water which support prolific booms of phytoplankton and other sea life.

These upwelling zones are of vital economic importance to the commercial fishing industry, and ship captains spend a great deal of time searching for upwelling currents. There are, however, fluctuations in the atmospheric and oceanic circulation systems that can cause periodic shifts of the coastal upwelling zones with potentially disastrous consequences for the world's food supply.

STORMS IN THE ABYSS

On the western side of the ocean basins, periodic undersea storms skirt the foot of the continental rise and transport huge loads of fine sediment, dramatically modifying the seafloor. The storms scour the seafloor in some areas and deposit large volumes of silt and clay in others. The energetic currents move about 1 mile per hour and can scour the ocean floor just as efficiently as a gale with winds up to 45 miles per hour can scour shallow areas near shore.

These dynamic events in the deep sea are similar to the general circulation of the atmosphere, which gives rise to weather systems around the world. The mid-latitude eddies that are often associated with strong winds and heavy precipitation

TABLE 7-1. History of the Deep Circulation in the Ocean.

AGE	EVENT
>50 million years ago	The ocean could flow freely around the world at the equator. There is a rather uniform climate and warm ocean even near the poles. The deep water in the ocean is much warmer than it is today. There are only alpine glaciers on Antarctica.
35-40 million years ago	The equatorial seaway begins to close. There is a sharp cooling of the surface and of the deep water in the south. The Antarctic glaciers reach the sea, with glacial debris in the sea. The seaway between Australia and Antarctica opens. Cooler bottom water flows north and flushes the ocean. The snow limit drops sharply.
25-35 million years ago	A stable situation exists with possible partial circulation around Antarctica. The equatorial circulation is interrupted between the Mediterranean Sea and the Far East.
25 million years ago	The Drake Passage between South American and Antarctica begins to open.
15 million years ago	The Drake Passage is open; the circum-Antarctic current is formed. Major sea ice forms around Antarctica, which is glaciated, making it the first major glaciation of the Modern Ice Age. The Antarctic bottom water forms. The snow limit rises.
3-5 million years ago	Arctic glaciation begins.
2 million years ago	An Ice Age overwhelms the Northern Hemisphere.

have their counterpart in energetic deep-sea eddies. The storms seem to follow certain well-traveled paths, as indicated by long furrows of sediment on the ocean floor (FIG. 7-5). The scouring of the sea bed and deposition of thick layers of fine sediment results in much more complex marine geology than what would develop simply from a rain of sediments. The periodic transport of sediment creates layered sequences that look similar to those created by strong wind storms found in shallow seas.

The greatest volume of silt and mud and the strongest bottom currents are found in the high latitudes of the western side of both the North and South Atlantic Ocean. These areas have the greatest potential for abyssal storms that ultimately form and shape the seafloor. They also have the largest drifts of sediment on Earth, covering an area over 100 miles wide, 600 miles long, and over 1 mile thick.

The entire continental rise off North and South America has been shaped largely by abyssal currents at depths from 2 to 3 miles. Elsewhere in the world, bottom currents have shaped the distribution of fine-grained material along the edges of Africa, Antarctica, Australia, New Zealand, and India.

Not all drifts are directly attributable to abyssal storms, and vast areas of the ocean have been modified by material carried by deep currents. The main role of the storms is to stir up sediment which is then picked up and carried downstream for long distances by the bottom currents.

During abyssal storms, the velocity of the bottom currents (FIG. 7-6) can increase from about 0.1 to 1 knot or more. (A knot is one nautical mile, or 1.15 statute miles, per hour.) This might seem rather slow compared to terrestrial storms, but water is a much denser medium than air.

The storms in the Atlantic seem to derive their energy from surface eddies shed from the Gulf Stream. While the storm is in progress, the suspended sediment load increases tenfold. The moving clouds of suspended sediment appear as coherent patches of turbid water with a lifetime of about 20 minutes. The storm itself might last anywhere from several days to a few weeks, at the end of which time, the current velocity drops back to normal and the sediment drops out of suspension.

The circulation of the deep ocean does not show a strong seasonal pattern like the circulation of the atmosphere. Therefore, the onset of abyssal storms are more unpredictable than the arrival of storms in the atmosphere, and they are likely to come to an area every 2 or 3 months.

LIFE AT THE TOP

The single most important aspect of the ocean is not its great depth or its vast volume, but its surface which, as a result of mixing by surface and upwelling currents and wind action, happens to be the most homogenous environment on Earth. Through the ocean's 140 million square miles of surface passes 70 percent of the sunlight, most of the rain, large amounts of oxygen and carbon dioxide, huge quantities of particulate matter, and alarming volumes of man-made pollutants. Almost half the area of the world's oceans is impoverished and is thought to account for only about 20 percent of the organic matter created through photosynthesis.

Phytoplankton, which live in the phototropic zone of the oceans where photosynthesis takes place, are called *primary producers* and are responsible for 80 percent of all oxygen production in the world. They also provide the basic food stock upon which all marine life ultimately depends for its survival. Any interference by man with the delicate balance within this thin membrane of water can have dire consequences for the ocean as well as the entire world.

The surface action of the ocean plays an important role in absorbing carbon dioxide from the atmosphere and mixing it with seawater (FIG. 7-7).

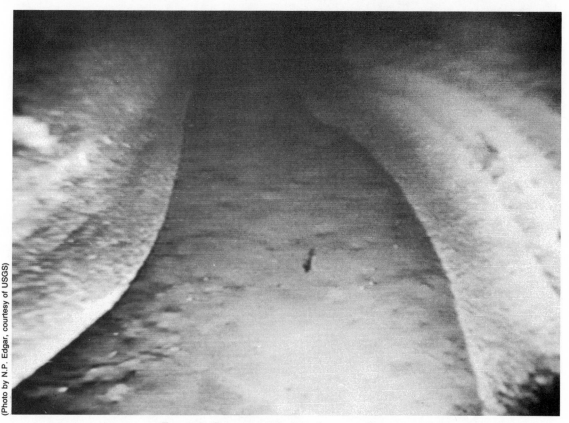

(Photo by N.P. Edgar, courtesy of USGS)

FIG. 7-5. Furrows in sand on the ocean floor.

If future projections are correct, the amount of atmospheric carbon dioxide will double from 0.03 percent to 0.06 percent sometime after the turn of the century. Most of this increase will be a result of the burning of fossil fuels and the destruction of forests.

The rate of carbon dioxide transferred across the ocean surface varies and is mostly dependent on small ripples, which have the effect of thinning the surface tension layer through which the gas must pass. Large waves also help stir in carbon dioxide, but they are only about half as effective. Simple plants that live near the surface absorb carbon dioxide during photosynthesis, and tiny animals also help absorption by stirring the surface layer. Turbulence in both the air and the sea are also important factors in the mixing of the ocean waters.

The wind-stirred layer of the ocean, which involves the top 300 feet, is in equilibrium with the atmosphere at all times. It can neither absorb more carbon dioxide, nor pass it on to deeper water because there is very little vertical mixing between layers in the ocean. The deep water, which

FIG. 7-6. Instrument to measure water currents on the seafloor.

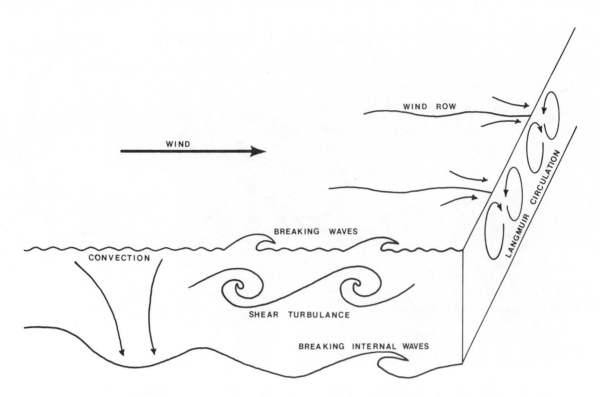

FIG. 7-7. Turbulence in the upper layers of the ocean induces mixing of temperature and nutrients.

represents about 90 percent of the ocean's volume, circulates very slowly and has a residence time on the order of about a 1000 years. It communicates directly with the atmosphere only in the polar regions, so its absorption of carbon dioxide is very limited in this manner. The abyssal receives most of its carbon dioxide in the form of shells of dead organisms and fecal matter which sink to the bottom.

Hydrocarbon chains, called *surfactants* (a contraction of the words *surf*ace-*act*ive) coat the surface of the ocean with a thin film that interferes with the transfer rates of gas and water vapor between the ocean and the atmosphere. An example is ordinary soap, which is a "dry" surfactant because it mostly rides on the surface of the water. This type of surfactant is rare, except in areas of man-made pollution, especially oil spills (FIG. 7-8), which produce a thin, suffocating film that rides on the surface. The more common variety are "wet" surfactants, which stick to the undersurface of the water. Although wet surfactants also retard all forms

of transfer at the surface of the ocean, they provide a mechanism for trapping and concentrating important substances that might otherwise escape from the ocean's surface. Thus, the ratio of dry to wet surfactants might be an important indicator of the health of the ocean.

Another type of transfer mechanism from the sea to the air involves the transport of marine substances by the wind. If simple evaporation of seawater were all that was involved, then rainwater would be almost as pure as distilled water. However, all forms of precipitation also contain minute amounts of salt and other materials derived from the sea. It appears that some fractionation process is taking place on the surface of the ocean, involving material derived from surface films. These substances are ejected into the air by the fine spray from bursting bubbles (FIG. 7-9) and from the ocean spray caused by the action of wind and waves. The spray evaporates into minute particles of sea salt that are wafted aloft by air currents.

FIG. 7-8. Transportation of oil on the high seas.

Upwards of 10 billion tons of salt are ejected into the atmosphere in this manner annually. The salt contains the principal metallic ions in seawater—namely sodium, calcium, magnesium, and potassium—along with other substances such as chlorine, bromine, iodine, phosphorus, and organic material. The salt particles play an important role in providing seed crystals, upon which raindrops form in clouds. They also deliver significant amounts of plant nutrients to coastal areas and enough food to sup-

port life in the high latitudes above the vegetation line. During a red tide, the organic material coughed up by the sea can include enough toxin to cause sore throats and eye irritation among shore dwellers.

Perhaps as much as 3 to 4 percent of the sea surface is covered with bubbles at any one time. Because very few marine organisms can extract dissolved nutrients directly from seawater, the particle-forming process by bursting bubbles is important in making phosphates and nitrates available to phytoplankton. The spray from bursting bubbles also might be involved in the precipitation of calcium carbonate, which causes the seawater to become iridescent with fine particles. This phenomenon, known as *whiting*, often occurs off the Bahamas. Each drop of spray from a bursting bubble also carries a positive charge of static electricity. Over the world's oceans, this amounts to a large, steady upward-flowing current.

A NAUGHTY LITTLE CHILD

For over a century, fishermen have used the Spanish term *El Niño*, which means "Christ Child," to denote the annual appearance of a warm-water

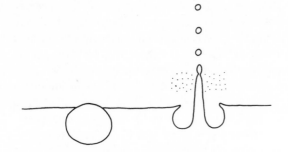

FIG. 7-9. Formation of particles of sea salt from bursting air bubbles.

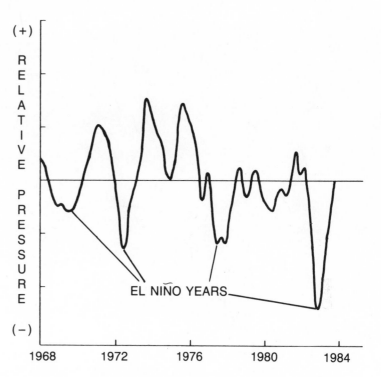

FIG. 7-10. The Southern Oscillation between Tahiti and Darwin, Australia.

EL NIÑO YEARS

1968   1972   1976   1980   1984

current off the coast of Ecuador and northern Peru that begins around Christmas. Normally the ocean surface in this area is quite cool compared with typical equatorial water, and it is kept that way by the Peru current, which carries cold water northward from the antarctic. The cold water is rich in nutrients that support a large Peruvian anchovy industry, the largest fishery in the world. Toward the end of December, a warm southward-flowing current from the equator displaces the cool water, reduces the upwelling of nutrients, and disrupts the fishing. The disruption is not widespread, nor does it last for very long. The warming usually ends by Easter.

Sometimes, however, an El Niño event can be much more intense, extensive, and prolonged. The surface-water temperature can be 10 to 15 degrees Fahrenheit above normal in places. The warming can spread all along the coast of Peru and the eastern and central equatorial Pacific, and the temperatures can stay high for a year or more. Intense El Niños have been observed in 1953, 1957-58, 1965, 1972-73, 1976-77, 1982-83 and 1986-87.

The El Niño warming event of 1972-73 was particularly devastating to the anchovy fishery, which

still has not recovered. The annual catch fell from a peak of more than 13 million tons in 1970 to less than 0.5 million tons in 1983. The disaster was so unprecedented that the name El Niño has since been applied, in scientific terminology, to these intense events, rather than to the much milder annual warming trends when the sea-surface temperature rarely exceeds 3 to 4 degrees Fahrenheit above normal.

The anomalous warming of the sea is linked with atmospheric pressure changes over the South Pacific, known as the Southern Oscillation (FIG. 7-10). When the pressure rises on Easter Island in the eastern Pacific, it falls in Darwin, Australia, in the western Pacific, and vice versa. When a major El Niño occurs, the barometric pressure over the eastern Pacific falls, while the pressure over the western Pacific rises. When El Niño ends, the pressure difference between these two areas swings in the opposite direction, creating a massive seesaw effect of atmospheric pressure.

The difference in atmospheric pressure between these two regions is known as the *Southern Oscillation Index*, which is used as a tool for predicting El Niño. When the index is unusually low, there is

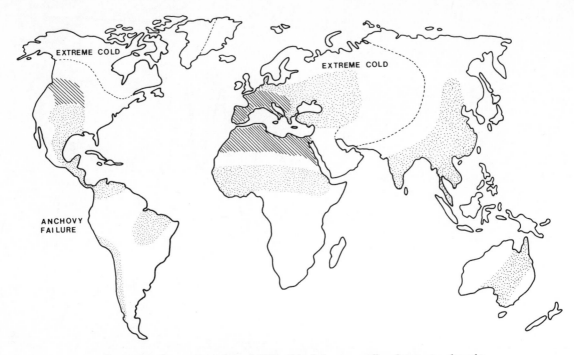

FIG. 7-11. Strange weather of 1972. Stippled areas suffered extreme drought.
Hashed areas had unusually wet weather.

a good chance for the occurrence of an El Niño, and a concurrent failure of the summer monsoon rains in India. The 1972-73 El Niño that devastated the Peruvian anchovy fishery came at a time when the index had fallen to one of its lowest values ever. It was also accompanied by a severe drought in India, while Peru was hit with heavy floods (FIG. 7-11).

Along the coast of South America, the southeast trade winds drive the Peru Current, pushing surface water offshore and allowing cold, nutrient-rich water to well up to the surface. The westward push of the trade winds continues across the eastern and central Pacific, and the resulting stress on the sea surface piles up water in the western Pacific. The warm surface layer of the ocean thickens in the west and thins in the east. In effect, this action lowers the boundary between the cold and warm layers of the ocean, called the *thermocline*, to about 600 feet in the western Pacific and raises it to about 150 feet in the eastern Pacific. The upwelling waters off the coast of South America are cold because the thermocline is so close to the surface.

During an El Niño event, which begins around October, there is a collapse of the trade winds in the western Pacific. With the warm water piled up in the western Pacific no longer supported by the winds, it flows back toward the east in the form of subsurface waves known as *Kelvin waves*, which can reach the coast of South America in 2 to 3 months, creating a great sloshing of seawater in the south Pacific basin.

The Kelvin waves generate eastward-flowing currents that bring in warm water from the west. This process lowers the level of the thermocline, which prevents the upwelling of cool water from below. With an increase of warm water from the west and the suppression of cool water from below, the sea surface begins to warm considerably by December or January. As the El Niño continues to develop, the trade winds begin to blow from the west, which intensifies the Kelvin waves and further depresses the thermocline off South America.

The Peru Current, running northward along the west coast of South America, is not significantly

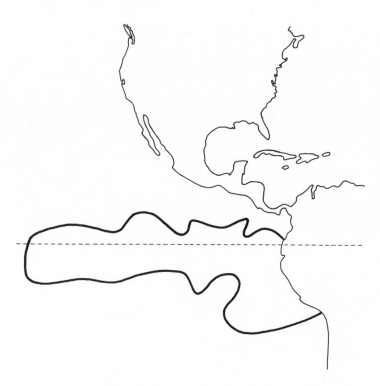

FIG. 7-12. Area affected by increased sea-surface temperature during the 1972 El Niño.

weakened by this action and continues to pump water to the surface, although this time, the upwelling water is warm and nutrient poor, rather than cold and nutrient rich. Also, the westward current off equatorial South America is not only weakened by the eastward push of the Kelvin waves, but is much warmer than before. This tends to spread the warming of the sea surface westward along the equator (FIG. 7-12). The normal wind pattern along the equator can reverse itself completely with the consequential reversal of weather patterns.

The El Niño of 1982-83 was remarkable for the amount of damage and human suffering it caused worldwide. Areas in the central and equatorial Pacific received copious rainfall. Day after day there was drenching rain, which disrupted the economy and the ecology. For example, 17 million birds on Christmas Island disappeared, never to be seen again. Meanwhile, much of Indonesia and eastern Australia was in the grip of record-breaking drought. In Indonesia, 340 people died of starvation. In Australia, because of the lack of food and water, thousands of hungry, thirsty cattle had to be shot and buried in mass

graves. Because of drought-related crop losses amounting to $2.5 billion, over 1 million people faced possible famine. Southern Africa suffered one of its worst droughts this century with water shortages, major crop failures, and widespread hunger.

Typhoons stalked the South Pacific. French Polynesia, which normally has one typhoon in 3 years was devastated by six major typhoons, while the Atlantic had its fewest (only two) hurricanes in 50 years. There were record floods in Ecuador and northern Peru, which took hundreds of lives, and unseasonably heavy rainfall continued along much of western South America. The California coast received destructive winds, tides, flooding, and landslides, which caused more than $300 million in property damages and forced 10,000 people to evacuate their homes. Storms marched across the South, placing portions of Mississippi, Louisiana, and Florida under water, and tens of thousands of people were forced to leave their homes (FIG. 7-13). The melting of an enormous snowpack in the Colorado Rockies brought the Colorado River to a record flood stage. There were severe dislocations of marine life

FIG. 7-13. Flooding from heavy rains in Denham Springs, Louisiana, April 7, 1983.

along the entire west coast of the Americas with a drastic drop in fish populations. In terms of economic losses and the amount of human misery it caused, the El Niño of 1982-83 could go down in the annals as possibly the single worst climatic event in modern history.

8

The Weather Machine

SCIENTISTS have only recently recognized the close association between the atmosphere and the ocean, and the ocean's large influence on the weather and ultimately the climate. The arrival of an El Niño is a prominent example of how elusive changes in the ocean and the atmosphere can sometimes lead to dramatic shifts in the behavior of the atmosphere, sending abnormal weather all around the world.

There are other quirks in the weather which, thanks to satellites, have been exposed within the last few years. One of these is a pulsation in the tropics that has a 40- to 50-day cycle and produces a wave of cloudiness in the Indian Ocean. As it sweeps eastward into the Pacific Ocean, it intensifies with speeds up to 20 miles per hour and then dies out when it reaches the eastern Pacific. As it circles the tropics, the oscillation can set parts of the atmosphere as far away as the poles pulsating at roughly the same frequency. This phenomenon apparently plays a roll in triggering the onset and withdrawal of the monsoons in India, causing the

rains to pause in midseason. The oscillation might play a major role in triggering El Niños. It also might effect the jet stream, which plays a central role in shaping the weather of North America.

THE GREAT HEAT ENGINE

Weather is one of the most complicated problems in all of science. It results in large measure from intricate interactions between the oceans and the *troposphere*, a blanket of air that extends 9 to 12 miles up from the Earth's surface (FIG. 8-1). The most striking feature about the troposphere is that it is in constant motion, and air masses move around the world on a grand scale. Warm air ascends from the equator, moves toward the poles, clashes with vast cold fronts, and produces storms. Hurricanes spin like colossal pinwheels 400 miles or more across, sucking up moisture from the oceans into their vortexes. All these forces produce everything from sunny skies over California to rain in Spain.

It has only been within the last decade that meteorologists have begun to understand that the

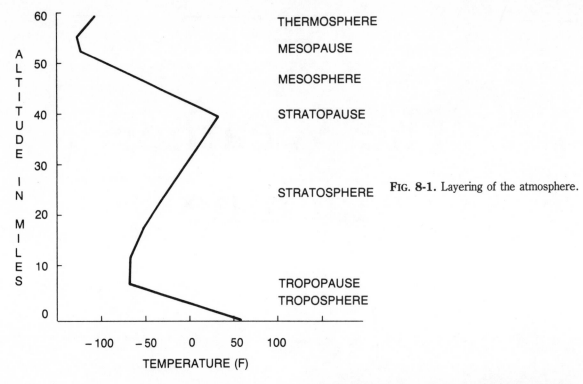

FIG. 8-1. Layering of the atmosphere.

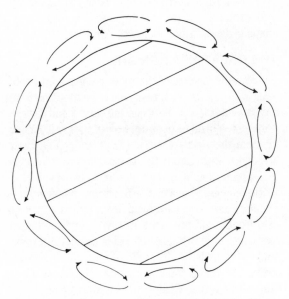

FIG. 8-2. Convection cells are responsible for distributing heat from the tropics to the poles.

atmosphere and the oceans act almost as a single fluid, exchanging both heat and gases. However, very little is known about what happens in the huge expanses of the ocean, particularly in the tropics where much of the nastiest weather forms.

In 1735, the English scientist George Hadley proposed his theory of heat convection. His thinking was that a current of air moved from the equator to the poles and back again in cellular structures called *Hadley cells*. When a parcel of air is heated at the equator, it rises to the *tropopause*, the boundary between the troposphere and the stratosphere. There it is blocked from climbing any further and is forced to move toward the poles. Because the Earth is spinning beneath it, however, the Coriolis effect bends the parcel of air to the east. When the parcel of air reaches the polar regions, it cools, sinks, and heads back toward the equator along a path that is bent to the west.

The Earth's rotation prevents the formation of whole equator-to-pole cells that Hadley envisioned. Instead, the circulation is accomplished by three separate cells in each hemisphere (FIG. 8-2), which

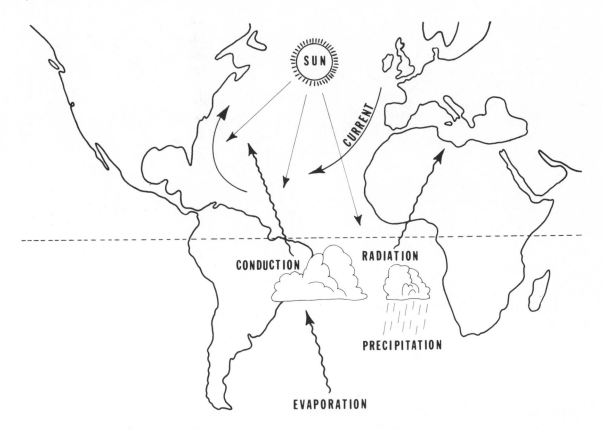

FIG. 8-3. Heat flow between the ocean and atmosphere.

transfer heat and cold to each other. These cells are responsible for the world's major wind belts including, from equator to pole: the doldrums, the horse latitudes, the trade winds, the prevailing westerlies, and the polar easterlies. The interaction between cells is made more complicated by the distribution of oceans, continents, mountain ranges, deserts, forests, and glaciers.

Because the greatest amount of heating takes place at the equator where the air is most directly under the Sun, the barometric pressure is lower there than it is in the zones on either side. Therefore, rising air is associated with low pressure and is created when air from both sides of the equator rushes in, meets head on, and rises.

Equatorial storms are formed when moisture is evaporated from the oceans and is carried toward the equator by the trade winds. Thus, the equatorial zone is also called the *Intertropical Convergence Zone,* or ITCZ. It varies in width from a few miles to about 60 miles, usually broader over the oceans, and its position changes from day to day, as well as from season to season.

The ITCZ is a major part of the thermal drive mechanism of the "heat engine" behind the weather (FIG. 8-3). When warm, moist air rises, it cools by expansion into lower pressures since air becomes thinner and colder with altitude. The cooling causes water vapor to condense into clouds. The condensation releases more thermal energy, causing the clouds to rise higher and higher with further cooling, which causes precipitation. The ITCZ is associated with some of the heaviest rainfall areas in the world, including the equatorial forests and jungles of South America and Asia. Its seasonal wanderings north and south of the equator, where it

FIG. 8-4. The major deserts.

draws in moisture-laden winds from the Indian Ocean, bring life-giving monsoons to southern Asia and parts of Africa.

The opposite action takes place in the subtropical zones where air is sinking, causing high pressure. After precipitation has been rung out of the rising air of the ITCZ, little moisture is carried by upper air currents to the subtropics. This produces what is known as the *horse latitudes*, so named because ships carrying horses to the New World were often becalmed in stifling heat for days on end. Many horses died and their carcasses were tossed overboard. The area is marked by generally clear skies and calm winds, and most of the world's great deserts exist under the sinking, dry air of the subtropics (FIG. 8-4). Also, some of the world's most popular seaside resorts are in these regions. As the high-pressure air of the horse latitudes subsides, it flows back toward the equator to complete the circulation of the tropical Hadley cell.

The intermediate Ferrel cells and the polar Hadley cells operate in much the same manner. Low pressure is accompanied by moist, rising air, and high pressure is accompanied by dry, sinking air.

The primary function of the general circulation of the atmosphere is to distribute heat from the tropics to the polar regions. If heat were not distributed, the tropics would be much too hot and the higher latitudes would be much too cold for human or animal habitation. The Earth intercepts about one-billionth of the Sun's energy, and only about half of this ever reaches the surface, where 90 percent of it goes to evaporate seawater. It takes a great deal of thermal energy to evaporate seawater in order to make clouds. When the clouds move to other parts of the world, they give up this energy in the form of precipitation, which effectively distributes the ocean's heat.

The general circulation is also coupled with the circulation of the ocean. The oceans take an active role by storing heat in the summer and slowly releasing it in the winter. The difference in temperature from day to night between the land and the ocean is also responsible for onshore and offshore breezes.

DATE	AREA OR HURRICANE	DAMAGE (IN $ MILLIONS)	DEATH TOLL
1881	Georgia and South Carolina		700
1893	Louisiana		2,000
1893	South Carolina		1,000-2,000
1900	Galveston		10,000
1913	Great Lakes		250
1919	Florida Keys and Texas		600-900
1928	Okeechobee		1,800
1935	Florida Keys		400
1938	New England	$ 300	600
1944	Atlantic Coast		400
1954	Carol	450	
1955	Diane	800	
1957	Audrey, Louisiana		400
1960	Donna	400	
1961	Carla	400	
1965	Betsy	1,400	
1969	Camille	1,400	250
1970	Celia	450	
1972	Agnes	2,100	
1975	Eloise	500	
1979	Frederic	2,300	5
1979	Claudette	400	
1979	David	300	
1983	Alicia	675	
1985	Elena	550	

The ocean's capacity for storing and moving vast quantities of heat surpasses that of the atmosphere by several fold. If the Sun went out, it would take several years before the oceans froze completely solid.

The advection of heat by ocean currents plays a major role in determining the climate. Ocean circulation is responsible for increased surface temperature at high latitudes, reduced snow and sea-ice coverage, and reduced sensitivity to daily and yearly changes in atmospheric carbon-dioxide concentrations. Thus, the moderating role of the oceans keeps most parts of the planet halfway between the freezing point and the boiling point of water.

WHIRLPOOLS IN THE SKY

In early September 1900, the United States suffered its worst natural disaster. A tropical storm born in the mid-Atlantic gathered strength as it moved westward, raking several Caribbean islands on its way to the Gulf of Mexico. Meanwhile, the citizens of Galveston, Texas, were unaware of this menace brewing in their backyard. When the hurricane began to bear down on Galveston, the seas quickly rose and submerged the only bridge connecting Galveston Island with the mainland, stranding most of the city's population of 38,000. As giant waves washed over the island, buildings crumbled, sending people into the surging waters, where they were taken helplessly out to sea. When it was all over, 10,000 to 12,000 people were dead—6,000 in the city of Galveston alone.

The world's worst storm-related catastrophe in modern history took place in the Bay of Bengal, Bangladesh, on November 13, 1970. Bangladesh is one of the world's poorest countries with a population of 100 million people crammed into an area about the size of Wisconsin. Millions of farmers have tried to eke out a living on low-lying islands called *chars* and along the coastal floodplains in the broad river delta of the Ganges and Brahmaputra rivers. An immense cyclone, spawned over the Indian Ocean, was drawn like a magnet up the Bay of Bengal (FIG. 8-5). Winds of over 100 miles per hour pushed up 50-foot tidal waves that surged into the bay and swept over islands and coastal plains. With no high ground to take refuge to, people simply tried to ride out the storm. Settlements were smashed and whole islands disappeared. When the storm passed, it left in its wake nearly 1 million dead.

Tropical cyclones are called *hurricanes* in the Atlantic, *typhoons* in the Pacific, *cyclones* in the Indian Ocean, and *willy-willies* in Australia. They are nature's most spectacular and most destructive storms. The energy released by a full-blown tropical cyclone is equivalent to a hydrogen bomb going off each minute during its lifetime.

The storms begin as tropical depressions over the oceans between latitudes 5 and 20 degrees north and south where the temperature of the sea surface is at least 80 degrees Fahrenheit. They must also be far enough away from the equator for the Coriolis effect to provide the necessary spin to create vigorous vortexes in the air. Therefore, tropical cyclones are more numerous in the summer and autumn when the Sun can heat the sea well to the north or to the south of the equator.

Only one out of ten tropical depressions actually grows into a tropical cyclone. Once they are formed, the storms are driven by the trade winds and normally head west (FIG. 8-6). This places the western regions of the oceans at the greatest risk. Sometimes a tropical cyclone will curve sharply away

Bay of Bengal Cyclone (HURRICANE) 1970
Photographed by ITOS-I

Nov. 11

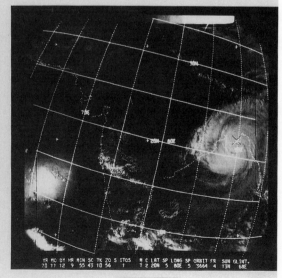
Nov. 12

(Courtesy of NOAA)

FIG. 8-5. The November 13, 1970, Bay of Bengal Cyclone.

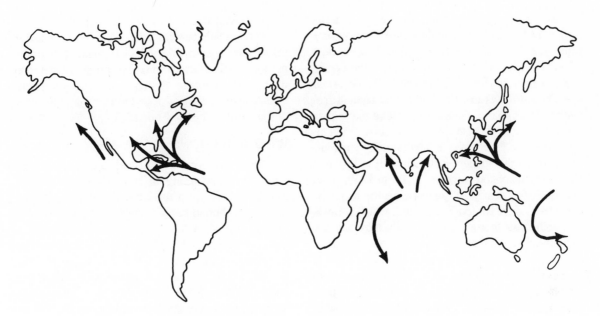

FIG. 8-6. Typical paths taken by tropical cyclones.

from the equator, enter colder water, and deprived of its source of heat energy, quickly dies away, becoming an ordinary depression again. Typically, most tropical cyclones have a life span of a week to ten days.

The majority of tropical cyclones are 300 to 400 miles across, but some can have diameters as much as 1200 miles and more. They rotate counterclockwise in the Northern Hemisphere and clockwise in the Southern Hemisphere. The circulating winds spiral in toward the center, or *eye*, at speeds of over 75 miles per hour. The eye ranges from 5 to 25 miles in diameter, and within the eye, the winds are fairly calm and the skies are nearly clear. The extremely low pressures in the eye suck seawater up into a huge mound several feet high. The winds pile up water in front of the tropical cyclone, which causes a tremendous storm surge or tidal wave that can wreck shore property and severely erode beachfronts.

Currently, the most destructive tropical cyclone to hit the United States was Hurricane Frederic, which came ashore near the Alabama-Mississippi line in mid-September 1979 and tore up $2.3 billion worth of real estate. Hurricane Camille, which attacked the

Mississippi Delta region on August 17, 1969, was the strongest and most terrifying hurricane in modern American history, killing 144 people and destroying $1.4 billion in property (FIG. 8-7).

When the tropical cyclone reaches land, it is deprived of its primary source of energy and must depend on heavy rainfall, which releases large amounts of latent heat, in order to stay alive. Therefore, tropical cyclones are an important source of rainfall throughout large parts of the world. Between 3 and 6 inches of rain is common, but upwards of several tens of inches is possible within a 24-hour period, which can cause severe flooding. As the tropical cyclone travels over land, the air flowing in at the bottom of the vortex is no longer moist, as it was over the ocean, and the tight structure of the vortex becomes wider and wider until the eye finally dissipates. The tropical depression can only regain its strength if it returns to the sea. Otherwise, it ends its days as just an ordinary low-pressure system.

Not all tropical cyclones even reach land, and in some cases, they seem to avoid islands. As the storm travels farther away from the equator, it encounters cooler water, and finally, it dies out and becomes a tropical depression when it reaches the higher latitudes. In the Atlantic, these low-pressure centers, or *cyclones*, are an important source of rainfall for Europe, and those that spawn in the Pacific march across the United States almost in procession.

The northwestern Pacific has the highest frequency of tropical cyclones, which take more lives and cause more damage than in any other region. Japan and the Philippines are particularly hard hit. More than 5000 Japanese lost their lives in Typhoon Vera, which flooded Nogoya, Japan, in 1959. In an average year, nearly 20 typhoons hit the major islands that comprise the Philippines, which sprawls across the main thoroughfare of the Pacific typhoons. In September 1970, a succession of typhoons struck the islands unusually hard, killing 1500 people. One of the worst disasters of the Pacific area was when a typhoon hit Hong Kong in 1937, killing 11,000 people. A similar storm in 1962 left 72,000 homeless, but luckily, the loss of life was not nearly as great.

It is with the loss of lives in mind that efforts have been made toward trying to tame these monstrous storms by cloud seeding. Unfortunately, tampering with these storms might just make them more dangerous, or worse, it might upset the worldwide heat-transport system with the consequential loss of desperately needed rainfall.

WILD GYRATIONS IN THE AIR

Thunderstorms are formed in an unstable air mass, generally during the summer and often along a cold front. A typical unstable atmosphere consists of a warm, humid surface layer overlain by a layer of cool, drier air. Often the two layers are separated by a thin inversion layer, in which the temperature increases with height. A parcel of air rising into this inversion layer is cooler than the surrounding air, and as a consequence, it is pushed back down. Thus, an inversion is very stable because it suppresses upward motion and confines instability to the lower

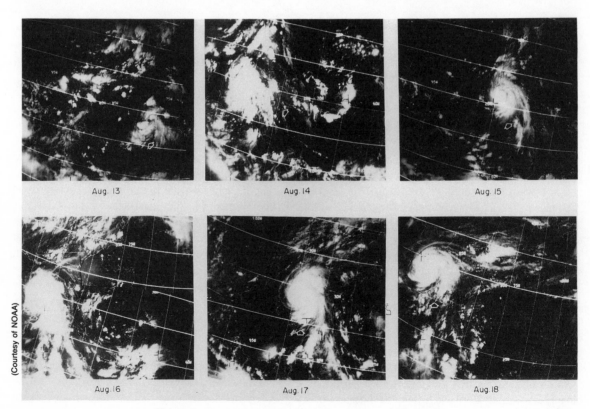

(Courtesy of NOAA)

FIG. 8-7. Track of Hurricane Camille, August 13-18, 1969.

FIG. 8-8. The jet stream.

layers. Eventually, as the day wears on and the ground heats up by the Sun, the air beneath the inversion layer becomes warmer, humid, and more unstable. It erodes the inversion layer or lifts it entirely, allowing the now highly unstable surface layer to erupt explosively, poking holes in the inversion layer at certain points. The surface air finds these leaks in the inversion layer, and tall thunderclouds form in the vigorous updrafts.

The jet steam also can weaken or completely dissipate a local inversion. The jet stream is a band of strong winds flowing between 5 and 7.5 miles altitude (FIG. 8-8). It is about 60 miles wide and about 0.5 mile deep. Wind speeds in the jet stream vary from 50 to 200 miles per hour, and 100-mile-per-hour winds are typical. The jet stream over the United States is usually associated with the *polar*

front, which is the boundary between polar and tropical air masses. The jet stream moves with the polar front farther to the north in the summer and farther to the south in the winter.

Within the jet stream, ribbons of highly strong winds several hundred miles long move downstream, pushing air down in front of them and drawing air up behind them. This uplifting action can dissipate an inversion layer and promote the formation of thunderstorms. If thunderstorms are already in progress, it can intensify them. Some of the storms might develop into hailstorms, which are particularly damaging to crops, and sometimes, they even herald the advent of a tornado. When a thunderstorm has the potential for developing tornadoes, this type of upper-air disturbance associated with the jet stream is almost always present. A tornado is,

TABLE 8-2. The Wind Scale.

BEAUFORT NUMBER	DESCRIPTION	MILES/ HOUR	INDICATIONS
0	Calm	< 1	Smoke rises vertically.
1	Light air	1-3	Direction of wind shown by smoke drift, but not by wind vane.
2	Light breeze	4-7	Wind felt on face; leaves rustle.
3	Gentle breeze	8-12	Leaves and small twigs in constant motion; wind extends light flag.
4	Moderate breeze	13-18	Raises dust and loose vapor; moves small branches.
5	Fresh breeze	19-24	Small trees in leaf begin to sway; crested wavelets form on inland water.
6	Strong breeze	25-31	Large branches in motion; telephone wires whistle.
7	Near gale	32-38	Whole trees in motion; resistance when walking against the wind.
8	Gale	39-46	Breaks twigs off trees; generally impedes progess.
9	Strong gale	47-54	Slight structural damage occurs.
10	Storm	55-63	Uproots trees; considerable structural damage occurs.
11	Violent storm	64-75	Widespread damage.
12-17	Hurricane	> 75	Devastation occurs.

therefore, a product of the interaction of a strong thunderstorm with the winds in the upper troposphere.

A strong updraft caused by a rising column of air in the heart of the thunderstorm begins to rotate counterclockwise and forms a tightly wound vortex of air several hundred feet in diameter (FIG. 8-10). The vortex only represents a small fraction of the tremendous energy of a thunderstorm, whose towering cumulonimbus cloud can reach 5 to 10 miles across and rise more than 10 miles high.

The vortex makes itself visible as a funnel cloud dangling part or all the way to the ground. A visible funnel cloud only forms if the pressure drop in the core reaches a critical value, which depends on the temperature and humidity of the inflowing air. Otherwise, if the funnel touches the ground, it takes on the color of the soil and debris it carries aloft. If a tornado spawns over the ocean, it sucks up large quantities of water and becomes a waterspout (FIG.

8-9). A typical tornado has a funnel that is usually cone-shaped, but very strong tornadoes can form broad pillars up to a mile across. Often there are ropelike tubes that trail off horizontally from the parent cloud.

During its brief lifetime, which is rarely more than a few hours, a tornado might abruptly change course and speed, loop around itself or around a twin or triplet tornado, hop about from place to place, completely disappear into the clouds, and reappear further downwind. Often tornadoes occur in wolf packs spread over hundreds of miles, and together, they cover large tracts of land, often traveling in a northeasterly direction at speeds of 30 to 60 miles per hour for 5 to 15 miles at a stretch. These tornado outbreaks are particularly dangerous, causing much damage and taking many lives.

Tornadoes can occur in other parts of the world, but by far the world's tornado hot spot with about 700 tornadoes yearly is the United States, Australia ranks second. The areas hardest hit are central and southeastern United States, including the states of Texas, Oklahoma, Arkansas, Kansas, Nebraska, and Missouri. The tornadoes occur in the spring and to a lesser extent in the fall when conditions are ripe for the formation of severe, tornado-producing thunderstorms. About three-quarters of the tornadoes reported in the United States occur between March and July. Tornadoes also accompany hurricanes when they make landfall during the summer and autumn, and cause much damage to coastal areas in the southeast.

Tornadoes can generate maximum wind speeds of 300 miles per hour (100 to 200 miles per hour is typical), making them the most violent weather phenomenon on earth. In these winds, loose objects become deadly missiles, and straw has been known to be driven through fence posts. The lifting power of tornadoes is tremendous. They can pick up locomotives and heavy trucks, and often, they carry livestock and people for long distances; but seldom does anyone ever make the trip alive.

Tornadoes take more lives in the United States than any other weather-related phenomenon, except for lightning which can strike suddenly without warning. The Tri-State Tornado Outbreak of March 18, 1925, was the worst ever, causing 689 deaths along a 200-mile path extending from southeastern Missouri to southwestern Indiana. The fierce tornadic winds also destroy buildings in the quarter-mile path

(Courtesy of NOAA)

FIG. 8-9. Waterspouts near Bahama Islands.

FIG. 8-10. The summer monsoons.

of destruction. Thus, about the only safe place to be during a tornado is in a cellar or preferably a storm shelter.

AS THE WIND BLOWS

Over 2 billion people on Earth depend on the monsoon rains for their agriculture. The term *monsoon* historically applies to the seasonal changes in wind direction on the shores of the Indian Ocean and the Arabian Sea, with winds blowing from the southwest during one-half of the year and the northeast during the other half. The word is derived from the Arabic *mausim* which means "season." Lately, however, the term has come to signify any annual climatic cycle with seasonal wind reversals that generally cause wet summers and dry winters.

During the rainy season, there are periods of drenching squalls interspersed with equal periods of a week or two of sunny weather. During the dormant phase of the monsoon, the weather is hot, dry, and stable with an absence of tropical storms. The largest and most vigorous monsoons are found on the continents of Asia, Africa, and Australia and the adjacent seas and oceans.

Two of the most important early studies on monsoons were done in the late seventeenth and early eighteenth centuries by the eminent English scientists Edmund Halley and George Hadley. Halley attributed the monsoon circulation primarily to the differential heating and cooling of the land and the ocean. This would cause pressure differences in the atmosphere that are equalized by the winds. Hadley noted that the rotation of the Earth would change the direction of the winds through the Coriolis effect, causing winds moving toward the equator to veer to the right in the Northern Hemisphere and to the left in the Southern Hemisphere.

More recent work has refined the understanding of these two processes, which are still considered the fundamental causes of monsoons. However, a third factor has been added to explain some of the more distinctive features of the monsoons. The ability of water to readily evaporate and condense in the atmosphere has a profound effect on monsoon circulation. Water evaporating from the world's oceans at any given time stores about one-sixth of the solar energy reaching the surface of the Earth. In monsoon circulation, part of this enormous reservoir of solar energy collected over the oceans is released over the land when the water in moist ocean air condenses. It is the release of this energy that is responsible for the power and the duration of the monsoon rainy season.

The *specific heat* of a substance is a measure of its ability to store heat. The specific heat of water is more than twice that of land. The temperature of dry land will, therefore, increase more than twice as much as the same mass of ocean to an equal amount of solar radiation. Since the Earth's surface is three-quarters water, the oceans absorb a great deal of heat. The greater heat capacity of the ocean is derived from its efficiency in mixing heat energy to the lower depths. By so doing, it distributes huge quantities of heat throughout a large mass of water. Turbulent eddies, which are stirred by wind blowing across the surface of the ocean, convey warm surface water to lower levels during the summer. In the winter, the heat accumulated during the summer is released when the surface water cools and is replaced by warmer water that rises from below. Because of the ocean's high specific heat and great mixing ability, the temperature of the ocean surface varies much less than that of the land. Moreover, the cycle of maximum and minimum surface temperature lags about 2 months behind the cycle of solar heating.

During the summer monsoon season, the difference in temperature between the land and the ocean increases the energy of the land-ocean system by setting up an atmospheric pressure difference between the two regions. The air over the ocean is cooler and therefore denser than the air over the land. In order to equalize the pressure differences, the cooler air from the ocean moves toward the land and undercuts the warm air over the land, forcing the warm air to rise. The combined rising of the warm air and sinking of the cool air releases a great deal of energy. This energy, combined with the steady input of solar energy, results in strong monsoon winds.

The circulation of the monsoon winds is deflected toward the east by the rotation of the Earth through the action of the Coriolis effect (FIG. 8-10). The summer monsoon will continue for as long as there remains unbalanced forces between the land and the ocean. As fall arrives, the temperature of the ocean drops, diminishing the temperature difference between the land and the ocean. The energy of the system runs down, the monsoon retreats, and the winter dry season begins. With the onset of winter, the land loses heat much faster than the ocean. The resulting increase heat loss from the land and the greater heat capacity of the ocean restores the energy of the land-ocean system, only this time the winds blow in the opposite direction.

When the summer monsoon is in progress, water vapor evaporated from the ocean is carried along with the wind blowing toward the land. A parcel of air carrying moisture from the sea is heated by conduction and by upward-moving convective air currents over the land, and it rises to higher altitudes. As the parcel ascends, it cools and the water vapor condenses into clouds, which further condense into rainfall.

The change of state of water from a vapor to a liquid releases large amounts of stored solar energy. The heat liberated in this manner adds considerable buoyancy to the rising column of air, and the parcel of air rises still higher, further reducing the pressure over the land and bringing in more moist air from the ocean. The release of energy in this way invigorates the monsoon circulation, which in turn increases rainfall.

It is this rainfall upon which the lives of nearly half the world depend. When the monsoons fail because of some climatic disturbance, millions of people starve. As the world's population continues to grow, a major drought caused by the failure of the monsoon winds could turn into the most tragic weather-related incident in human history.

The Water Cycle

WATER is a molecule composed of an oxygen atom and two hydrogen atoms, which are attached to the oxygen about 105 degrees apart. The hydrogen atoms have a slight positive charge, and the oxygen atom has a similar weak negative charge. These charges allow water molecules to attract one another and form in groups of up to eight molecules. More groups are formed near the freezing point than in warmer water, and because the groups take up more space, water expands upon freezing, thereby making ice less dense and able to float. Water reaches its maximum density at 4 degrees Celsius, thereby making it heavier and causing it to sink.

It requires 80 calories of heat to melt one gram (0.035 ounce) of ice, known as the *latent heat of melting*; 100 calories of heat to raise the temperature to the boiling point, called *sensible heat*; and 540 calories of heat to make water vapor, known as the *latent heat of vaporization* (FIG. 9-1). The conversion of water vapor back to one gram of ice liberates 540 + 100 + 80 = 720 calories of heat, which is a lot of energy, giving water a very high specific heat, or heat capacity.

Water is a poor conductor of electricity, but is an excellent solvent, able to dissolve most natural substances found on Earth. Other than metallic mercury, water is the only mineral that exists naturally on the Earth's surface as a liquid.

Three percent of the Earth's water is fresh, and there is enough freshwater to fill the Mediterranean Sea ten times over. About three-quarters of the freshwater is glacial or polar ice. The remaining freshwater, less than 1 percent of all water, is atmospheric water vapor, running water in rivers, standing water in lakes, groundwater, soil moisture, and water in plant and animal tissues. As far as terrestrial life is concerned, these are the most important sources of water. Without them, the entire surface of the planet would become a dry, barren wasteland.

Each day, the Sun evaporates a trillion tons of seawater. However, because the atmosphere can only hold about 0.5 percent of the total freshwater supply at any one time, all water evaporated from the oceans must fall to Earth as precipitation. Since 70 percent of the Earth's surface is ocean, most of

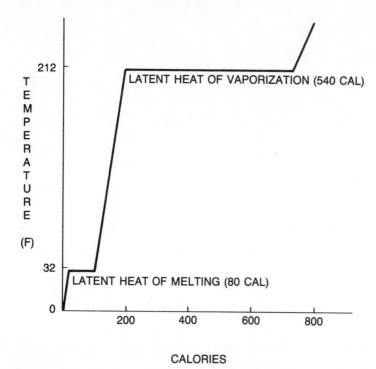

FIG. 9-1. The change of state of water.

this precipitation returns directly to the sea. The rest falls on the continents, where some of it is evaporated back into the atmosphere. The majority of the water on the land eventually makes its way back to the sea, completing the loop in one of nature's most important life processes, known as the *hydrologic cycle* (FIG. 9-2).

BALANCING THE BUDGET

The Earth must radiate back into space exactly the same amount of heat energy it receives from the Sun. If it did not emit enough heat energy, the surface of the Earth would get intolerably hot, and if it radiated too much, the Earth's surface would get intolerably cold. This delicate balancing act is known as the *heat budget* (FIG. 9-3).

The Earth receives only about one-billionth of the Sun's energy; yet, this is still several thousand times greater than all the energy that is now consumed by man. It is also about 100,000 times greater than the heat that reaches the surface from the Earth's molten interior. If solar energy averaged over a year were spread evenly around the surface

TABLE 9-1. Albedo of Various Surfaces.

SURFACE	PERCENT REFLECTED
Clouds, stratus	
<500 feet thick	25-63
500-1000 feet thick	45-75
1000-2000 feet thick	59-84
Average all types and thicknessess	50-55
Snow, fresh-fallen	80-90
Snow, old	45-70
White sand	30-60
Light soil (or desert)	25-30
Concrete	17-27
Plowed field, moist	14-17
Crops, green	5-25
Meadows, green	5-10
Forests, green	5-10
Dark soil	5-15
Road, blacktop	5-10
Water, depending upon sun angle	5-60

of the Earth, it would amount to about 300 watts per square yard, or about 1000 megawatts (million watts) per square mile. If this amount of solar energy were converted into electricity at 10 percent efficiency, there would be enough electrical power to run a fair-sized city.

Various surfaces have different *albedos*, which is the ability to reflect sunlight and is mostly dependent upon color. Light-colored objects reflect light, and dark-colored objects absorb it. Snowfields and deserts have a high albedo and reflect most of the sunlight, and the oceans and forests have a low albedo and absorb most of the sunlight.

The angle that sunlight impinges on the Earth's surface also determines how much solar energy is absorbed and how much is reflected. At the equator, the Sun's rays strike the surface directly, and more solar radiation is absorbed than is reflected. In the polar regions, the Sun's rays strike the surface at an indirect angle, and more solar radiation is reflected than is absorbed. This fact and their high albedo are two of the reasons the poles remain covered with ice all year round.

The sunlight striking the Earth is absorbed by the atmosphere, reflected back into space by clouds, and scattered sideways by dust particles and aero-

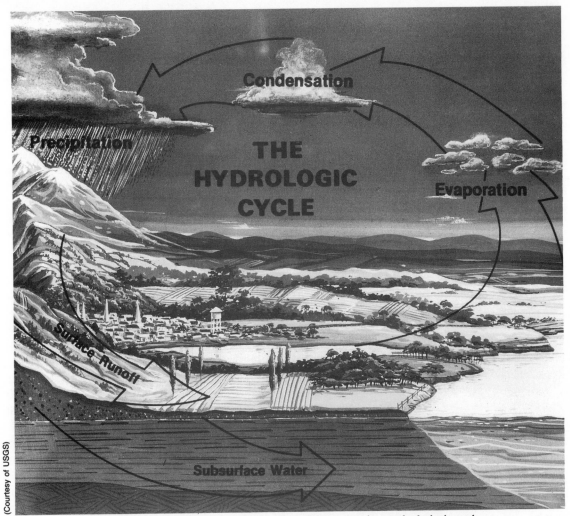

FIG. 9-2. All water from the ocean is returned in a continuous hydrologic cycle.

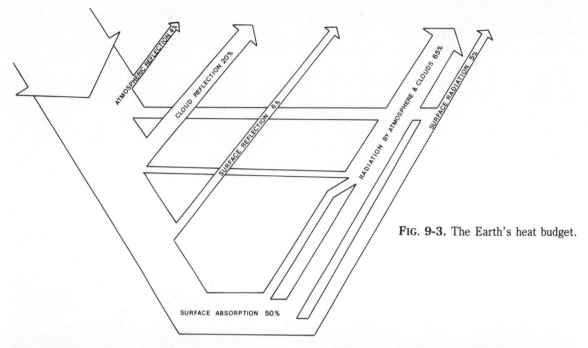

FIG. 9-3. The Earth's heat budget.

sols; only about half is allowed to pass on to the surface. More than 30 percent of the Sun's energy is reflected back into space before it ever has a chance to warm the Earth. Most of the sunlight that reaches the surface is absorbed by the oceans, where it is used to evaporate seawater.

When sunlight strikes the land, most of it is absorbed by soil and vegetation and converted to infrared radiation. Infrared radiation has a longer wavelength than visible light and is therefore invisible. However, its effects on the air just above the surface are made visible as waves of heat, which are responsible for making mirages. The infrared radiation is absorbed by the atmosphere or reflected to the surface by clouds, and eventually, it is radiated back out into space.

About 15 percent of the moisture in the atmosphere comes from the land by the evaporation of surface water and soil moisture, and the transpiration of plants. When water vapor distilled from the oceans and evaporation from the land condense into clouds and precipitate as rain or snow, large quantities of stored solar energy are liberated, much of which escapes into space, and the balance is maintained.

RUNNING OFF WITH THE RUNOFF

The continents lose some 15,000 cubic miles of water to evaporation each year, but they receive about 25,000 cubic miles of precipitation, averaging roughly 25 inches annually over the entire land surface (FIG. 9-4). That leaves a surplus of about 10,000 cubic miles of freshwater. Much of this is lost through floods or is held in the soil or in swamps. About one-third is *base flow*, or stable runoff of all the world's rivers and streams. Another third is *subsurface flow*, which discharges mostly through evaporation, with only about 1 percent reaching the oceans. Another 1000 cubic miles of freshwater flows into barren regions. That leaves less than 10 percent of all precipitation on the land for man's use.

The amount for man's use averages about 2000 tons, or about 0.5 million gallons, of freshwater per person per year for the entire world. That might seem like a lot of water, but only about 30 tons, or 7500 gallons, is actually used for domestic purposes, and of this amount, only 1 ton is used for drinking water. The vast majority of the available freshwater is used for agriculture and industry. The figures are also lopsided because, although rich nations represent only about 10 percent of the world's popula-

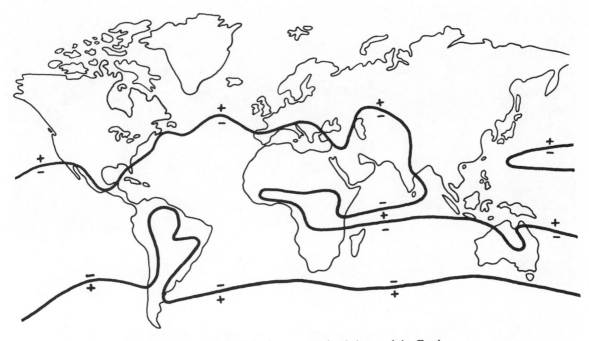

FIG. 9-4. The precipitation-evaporation balance of the Earth.
In positive areas, precipitation exceeds evaporation. In negative areas, evaporation exceeds precipitation.

tion, they consume and also waste most of the world's freshwater supply.

Over 10 percent of the world's cultivated land is irrigated, requiring 600 cubic miles of water (FIG. 9-5). The advantages of irrigation are that crops do not need to depend upon the whims of nature for their water, more land can be brought under cultivation, and two or more crops can be grown in a single year. Its disadvantages are that river water used for irrigation has a high salt content and if fields are not drained properly, the salt buildup could ruin the soil. Thousands of acres are rendered useless by this process yearly.

Also, overirrigation on poorly drained fields can waterlog the soil and ruin crops. Furthermore, chemicals such as fertilizers and pesticides are carried off with the drain water and end up in streams and eventually the ocean. The depletion of water for irrigation from rivers such as the Colorado and the Arkansas, which flow out of the Colorado Rockies and are now only a trickle of their former selves, have led to heated water disputes in the American West and Midwest.

Groundwater is the second most important freshwater resource in the world. An *aquifer* is composed of unconsolidated sand and gravel, and water flows through this formation by the influence of gravity at most only a few inches per year. Water catchment areas at the head of the aquifer recharge the groundwater system. The rate of infiltration into the groundwater system depends upon the distribution and amount of precipitation, the type of soils and rocks, the slope of the land, the amount and type of vegetation, and the amount of water rejected because the soil is already saturated.

The use of groundwater for irrigation is expensive, and only affluent nations can afford to use it on a large scale. The overuse of aquifers has led to the lowering of the water table or depletion of aquifer altogether. Once an aquifer is depleted, it cannot be restored to its original capacity because subsidence compresses the sediments, which in turn decreases the pore spaces between grains where water flows.

Industry is the second heaviest user of freshwater. Industrial demand in the United States is about 0.5 million gallons per person per year, which

FIG. 9-5. River water irrigation.

is about 100 times greater than that of the developing countries. The manufacture of a single automobile requires tens of thousands of gallons of water, beginning with the mining, milling, and smelting of steel; the manufacturing of glass, rubber, and plastics; the casting, boring, die stamping, and making of various subassemblies; and its final bath before going into the showroom. Other industries that rely heavily on water include petroleum refining, chemical manufacturing, pulp and paper manufacturing, and food processing.

Unfortunately, much of the industrial use of water involves the dissolving of toxic chemicals, which end up in rivers on their way to the sea. Also, coolant water returned to lakes and rivers from electrical generating stations increases the water temperature significantly enough to disturb aquatic ecosystems.

An abundant source of freshwater that has been largely overlooked is icebergs. Every year some 5000 icebergs, totaling about 200 cubic miles of ice, are calved from Antarctica. Each iceberg has a flat top and steep sides, and consists of approximately 100 million tons of freshwater. The iceberg drifts along with the wind and the currents for a number of years, finally breaking up and melting when it enters warmer water. Occasionally, an iceberg will drift as far north as 30 degrees south latitude, and might influence the regional climate, as well as the local weather.

Studies have been devoted to the feasibility of capturing icebergs and towing them to Australia, South Africa, Saudi Arabia, and even California and utilizing them as a supply of freshwater. Only very large icebergs could be used in this manner because smaller ones have a tendency to break up even in moderate seas. Also, melting and erosion is a major factor, and the larger the iceberg, the better are the chances of a substantial block of ice making it to its destination. Instead of the iceberg being melted upon arrival, it would be mined, using existing technology. A slurry of ice and water then would be pumped ashore via an ocean pipeline. However, such a scheme is not without its dangers, not only because of accidents resulting from broken off icebergs in the shipping lanes, but also from the uncertainty over the environmental effects of moving large numbers of icebergs around the world.

BUSY AS A BEAVER

Water management entails the construction of a large number of dams for hydroelectric power, river navigation, irrigation, city reservoirs, fisheries, recreation, and, most important of all, flood control (FIG. 9-6). By controlling floods, a dam can make some of the world's most fertile land in the floodplains available for safe habitation and cultivation. Thus, by capturing floodwaters that would otherwise flow into the sea, a dam can augment the total water supply.

The generation of hydroelectric power is an industrial use of water that is entirely nonconsumptive because the water must be released to other water users downstream anyway. In the past, hydroelectric power generation was so economically appealing that almost all suitable sites have already been exploited. For the remaining sites, the question now is whether the energy extracted and the economic benefits of irrigation and other uses can outweigh the costs of constructing large dams.

The capital expenditure for a large dam is tremendous, roughly $0.5 billion for each cubic mile of capacity, and the costs keep rising. The damming of large bodies of water inundates huge tracks of land, while enlarging the surface area of the reservoir, thereby increasing the loss of water through evaporation. The life expectancy of a dam is not indefinite, and eventually the entire reservoir will fill up with silt. The most effective means of controlling silt buildup is by adopting soil-conservation measures in the watershed so that less topsoil is lost to erosion.

China has solved some of these problems by building huge numbers of small dams, thereby reducing the cost of construction, which can be done by the local labor force. Some 100,000 dams and reservoirs in China either completed or still under construction will have a total storage capacity of about 100 cubic miles of water. This will give China the largest volume of regulated stream flow, most of it for irrigation, of any country in the world.

FIG. 9-6. Hoover Dam and Lake Mead on the border between Nevada and Arizona.

There is a reason why most of the world's major cities originally were located near watercourses. Since the Neolithic period, beginning about the time when the glaciers started to melt some 10,000 years ago, human settlements have been clustered in the major river basins precisely because water was readily available there. Rivers became the main means of transporting goods, and boats plied their way from one prosperous river port to the next.

Modern navigation of inland waterways makes a large demand on water resources. A dam is built to maintain a high water level behind the dam as well as below it, requiring the release of large volumes of water. Nevertheless, river valleys will remain the primary focus of human civilization and continue to supply both fertile land and water resources. Moreover, because rivers offer abundant water for consumption and navigation, they will continue to be the most logical sites for major industrial development.

The two rivers with the largest annual flow—the Amazon with 1300 cubic miles and the Congo with 300 cubic miles—are also among the least exploited, mainly because they flow through inhospitable rain forests. Large rivers that drain into the Arctic Ocean, such as the Mackenzie in Canada and the Ob and Yenisei in the Soviet Union, are also virtually unutilized. This might change if the Soviets divert the Ob and Yenisei rivers across warmer and more arable land through a series of giant dams and canals to the Aral Sea to the southwest. A similar proposal calls for the diversion of the Mackenzie River from its present channel across the Canadian Arctic. It is not yet known what effect such large-scale intervention in the water cycle might have on the arctic environment and on the global climate.

At the opposite extreme is the Nile, which has an annual flow of only 20 cubic miles, yet it is completely developed. Since the mid 1970s when Lake Nasser behind the Aswan High Dam was filled, the Nile has practically ceased to flow into the Mediterranean Sea. Instead, its waters now serve one of the largest irrigated areas in the world, amounting to some 20,000 square miles, or about the size of West Virginia. Elsewhere in Africa, on the border between Zambia and Zimbabwe, the Zambezi River has been dammed to create Kariba Lake the largest artificial reservoir in the world. Africa, which is prone to drought, depends heavily on its water projects for irrigation.

All the large rivers in southern and southeastern Asia such as the Yangtze, the Mekong, the Irrawaddy, the Brahmaputra, the Ganges, and the Indus also have been extensively developed. This gives Asia the largest volume of impounded water, mostly for irrigation, of any other continent.

TOO MUCH OF A GOOD THING

Floods are essentially a man-made disaster. They are natural, recurring events and only become a hazard when man builds on the floodplain. The purpose of the floodplain is to carry away excess water during a flood, and man's failure to recognize this function has led to haphazard development in these areas with a consequent increase in flood hazards. Floodplains provide level ground, fertile soils, ease of access, and available water supplies, but because of economic pressures, they are being developed without full consideration of the flood risk. As a consequence, the American government has stepped in and assumed much of the responsibility for providing flood relief. In addition, the government has spent more than $9 billion on flood-protection works since 1936.

Despite these programs, however, the average annual flood hazard has been on the increase because people have been moving into flood-prone areas faster than flood-protection projects are being constructed. Therefore, the increased losses are not a result of greater floods, but of increased encroachment onto the floodplains. As the population increases, there is more pressure to develop flood-prone areas without taking the proper precautions. When their property is flooded, people expect the government to bail them out.

There are about 3 million miles of rivers and streams in the contiguous United States, and about 6 percent of the land area is prone to flooding (Fig. 9-7). A large percentage of the nation's population and property is concentrated in these flood-prone areas. More than 20,000 communities have flood problems, and of these, about 6000 have populations greater than 2500.

TABLE 9-2. Chronology of Major U.S. Floods.

DATE	RIVERS OR BASINS	DAMAGE (IN $ MILLIONS)	DEATH TOLL
1903	Kansas, Missouri and Mississippi	$ 40	100
1913	Ohio	150	470
1913	Texas	10	180
1921	Arkansas River	25	120
1921	Texas	20	220
1927	Mississippi River	280	300
1935	Republican & Kansas	20	110
1936	Northeast U.S.	270	110
1937	Ohio & Mississippi	420	140
1938	New England	40	600
1943	Ohio, Mississippi, and Arkansas	170	60
1948	Columbia	100	75
1951	Kansas & Missouri	900	60
1952	Red River	200	10
1955	Northeast U.S.	700	200
1955	Pacific Coast	150	60
1957	Central U.S.	100	20
1964	Pacific Coast	400	40
1965	Mississippi, Missouri and Red Rivers	180	20
1965	South Platte	400	20
1968	New Jersey	160	- - -
1969	California	400	20
1969	Midwest	150	- - -
1969	James	120	150
1971	New Jersey & Pennsylvania	140	- - -
1972	Black Hills, S. Dakota	160	240
1972	Eastern U.S.	4,000	100
1973	Mississippi	1,150	30
1975	Red River	270	- - -
1975	New York & Pennsylvania	300	10
1976	Big Thompson Canyon	- - -	140
1977	Kentucky	400	20
1977	Johnstown, Pennsylvania	200	75
1978	Los Angeles	100	20
1978	Pearl River	1,000	15
1979	Texas	1,250	- - -
1980	Arizona & California	500	40
1980	Cowlitz, Washington	2,000	- - -
1982	Southern California	500	- - -
1982	Utah	300	- - -
1983	Southeast U.S.	600	20

FIG. 9-7. Flood-hazardous areas in the United States.

Floods threaten life, cause much suffering, heavily damage property, wipe out crops, and halt commerce. The average annual flood loss in the United States has increased from less than $100,000 at the beginning of the century to more than $3 billion today. By the year 2000, if trends continue, the potential annual flood loss is expected to be more than $4 billion. Presently, some 100 lives are lost annually in the United States because of flooding. From 1947 to 1967, over 150,000 lives were lost from flooding in southern and southeastern Asia. In that same period, about 1300 lives were lost in the United States.

Flash floods are the most intense form of flooding. They are local floods of great volume and short duration, and generally result from a torrential rain or cloudburst associated with severe thunderstorms on a relatively small drainage area. Flash floods also result from a dam break or from a sudden breakup of an ice jam, each causing the release of a large volume of flow in a short time. A special type of flash flood occurred during the 1980 eruption of Mount St. Helens, which produced major mudflows and flooding from melted glaciers and snow on the volcano's flanks.

Flash floods from violent thunderstorms produce flooding on widely dispersed streams, resulting in high flood waves. The discharges quickly reach a maximum and diminish almost as rapidly. Floodwaters frequently contain large quantities of sediment and debris which are collected as the waters sweep clean the stream channel.

Flash floods can take place in almost any part of the country, but they are especially common in the mountainous areas and desert regions of the West. They are particularly dangerous in areas where the terrain is steep, surface runoff rates are high, streams flow in narrow canyons, and severe thunderstorms are prevalent.

Riverine floods (FIG. 9-8) are caused by precipitation over large areas, by the melting of the winter's accumulation of snow, or both. They differ from flash floods in both extent and duration, and they take place in river systems whose tributaries drain large geographical areas and encompass many independent river basins. Floods on large river systems might last from a few hours to many days. The flooding is influenced primarily by variations in the intensity and the amount and distribution of precipitation. Other factors that directly affect flood runoff are the condition of the ground, the amount of soil moisture, the vegetative cover, and the amount of urbanization where the ground is covered by impervious pavement.

River channel storage, changing channel capacity, and timing of flood waves control the movement of floodwaters. As the flood moves down the river system, temporary storage in the channel reduces the flood peak. As tributaries enter the main stream, the river becomes larger downstream. Since tributaries are not the same size nor are they spaced uniformly, their flood peaks reach the main stream at different times, thereby smoothing out the peaks as the flood wave moves downstream.

Tidal floods are overflows on coastal lands bordering the ocean, an estuary, or a large lake. These coastal lands, including bars, spits, and deltas, are affected by the coastal current and offer the same protection from the sea that floodplains do from rivers. Coastal flooding is primarily a result of high tides, waves from high winds, storm surges, seismic seawaves from submarine earthquakes, or any

(Courtesy of U.S. Army Corps of Engineers)

FIG. 9-8. An overtopped back levee in New Orleans, Louisiana.

combination of these. Tidal floods also are caused by the combination of waves generated by hurricane winds and flood runoff resulting from the heavy rains that accompany hurricanes.

Flooding can extend over large distances along a coastline. The duration is usually short and is dependent upon the elevation of the tide, which usually rises and falls twice daily. If the tide is in, then other forces that produce high waves can raise the maximum level of the prevailing high tide. The most severe tidal floods are caused by tidal waves generated by high winds superimposed on regular tides. Hurricanes are the primary sources of extreme winds, and each year several of these storms enter the American mainland.

TREATING THE SOIL LIKE DIRT

Efforts to increase worldwide crop production through the use of irrigation, fertilizers, genetic en-

gineering, and improved farming techniques all will be thwarted if the topsoil is eroded away. This problem becomes increasingly alarming as human populations continue to grow. In the United States, eroding cropland is costing the country nearly $1 billion each year in polluted and sedimented rivers and lakes. The Soil Conservation Service estimated that in 1977 alone, about 3 billion tons of soil from cultivated fields either washed or blew away. That would make a mound of dirt roughly 1 mile on a side and 1 mile high.

Part of the problem is that, during the decade of the seventies, American farmers began to export massive quantities of grain to feed a hungry world; most of it went to the Soviet Union. An additional 60 million acres, a larger area than the entire state of Kansas, came under the plow to keep up with the additional demand. Some of this increased land area contained sloping, marginal, and fragile soils that were easily erodable (FIG. 9-9).

(Courtesy of USDA)

FIG. 9-9. Severe erosion from use of soils unsuitable for cultivation.

The decade of the 1980s saw higher competition from abroad, which resulted in dwindling markets for American farmers. Farmers who invested in high-priced land and expensive farm implements at exorbitant interest rates found no markets for their produce, and foreclosures were at epidemic levels. The situation was made even worse as farmers sacrificed their land in order to make a fast buck.

Erosion is a natural geologic process that cuts down mountains to make flat plains. It has been going on since the beginning of time. Erosion rates vary, depending on the amount of precipitation, the topography, the type of rock and soil, and the amount of vegetative cover. Each year, the Mississippi River pours more than 0.25 billion tons of sediment into the Gulf of Mexico, widening the Mississippi Delta and slowly building up Louisiana. The Gulf Coast States from east Texas to the Florida panhandle were built up from sediments eroded from the interior of the country and hauled in by the Mississippi and other rivers. The Imperial Valley of southern California owes its rich soil to the Colorado River, which carved out the mile-deep Grand Canyon and left sediments 3 miles thick on the valley floor.

Soil begins as ordinary rock. Weathering by wind and water peals off layer by layer in a process called *exfoliation*. The rock is then reduced to sand, silt, and microscopic clay particles. Most soils are about half mineral; the rest is air and water mixed with organic material, such as the remains of plants and animals.

The soil is divided into three major layers (FIG. 9-10). The A horizon is the topsoil, containing the maximum accumulation of organic matter and the maximum leaching of clay, iron, and aluminum. This is where seeds germinate and grow. The B horizon is the subsoil, which collects clay, iron, and aluminum leached from the A horizon. The C horizon is the parent rock from which new soil is formed.

Soil types vary by the type of parent rock, the climate, the topography, the type of organisms, and the length of time required to make new soil. In the United States, there are upwards of 30,000 different soils, but generally they can be broken down into

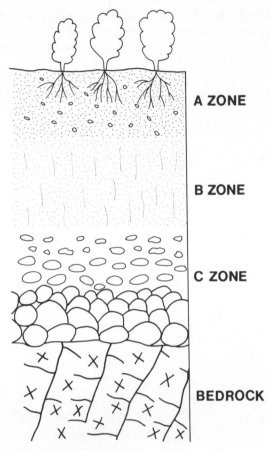

A ZONE

B ZONE

C ZONE

BEDROCK

FIG. 9-10. The soil profile: A-organic rich; B-organic poor; C-parent rock.

three basic soil types. *Pedalfers* are enriched in iron and aluminum and are found extensively in the East. *Pedocals* are enriched in calcium and are found extensively in the West. *Podzols* are enriched in organic matter and are found in forested areas, where the soil is generally thin and less suitable for cultivation.

Erosion is a serious problem on about half the nation's cropland. In many other parts of the world, in order to keep up with food demand, farmers have adapted farming practices that are leading to excessive rates of soil erosion. As the A horizon becomes thinner and erosion brings the B horizon closer to the surface, the potential for runoff and erosion is increased because the B horizon is generally unfavorable for plant root growth which tends to hold the soil in. The eroded soil is carried by tributaries, which empty into rivers, which empty into the sea, and ultimately the soil ends up on the ocean floor.

Once the soil is gone, only nature can put it back and only over a period that is measured on a geologic time scale. Before the advent of agriculture, which followed the retreat of the ice age some 10,000 years ago, natural soil-erosion rates were in equilibrium with the production of new soil. Present soil-erosion rates have more than doubled the rate that nature is putting the soil back.

10

Making Waves

A LITTLE over sixteen centuries ago, in A.D. 365, a tremendous earthquake that originated in the Mediterranean Sea leveled coastal outposts and created a tidal wave that destroyed the Egyptian port of Alexandria. The earthquake's epicenter was about 30 miles southwest of the ancient Roman coastal city of Curium in southern Cyprus. The earthquake was of such intensity that it shattered the walls of buildings and caused near total destruction. Similar types of destruction 30 miles away from Curium indicated that the earthquake must have been enormous and that aftershocks probably continued for years afterwards. On the modified Mercalli scale, the earthquake rated the highest score of 12.

In Curium, people had no warning and little or no time to get away when the earthquake hit. People and animals were caught in mid-stride as the walls came tumbling down. The entire city was buried in rubble, and a slice of Roman culture was preserved among the ruins. The high-intensity undersea earthquake caused the formation of huge seismic seawaves that pounded the shores of the Mediterranean for hundreds of miles around.

CATCH THE WAVE

Most waves in the ocean are generated by large storms at sea when the wind blows across the surface of the water. As long as the water depth is over one-half the wavelength, the waves are considered deep-water waves. If the wavelength is longer than the depth of the water, the waves are called shallow-water waves. When the waves approach the shore and the depth becomes less than one-half the wave length, the wave feels the bottom and slows down. This shoaling of the wave causes its shape to be distorted, and the wave breaks along the beach (FIG. 10-1). Wave breaking dissipates wave energy along the coast and is converted into the process of erosion.

Breaking waves are also responsible for generating along-shore currents, which transport sand along the beach. Wave reflection bounces wave energy off of steep beaches or seawalls and is responsible for the formation of sandbars. When waves approach the shore at an angle to the beach, the wave crests are bent by refraction. When waves pass the end of a point of land or the tip of a break-

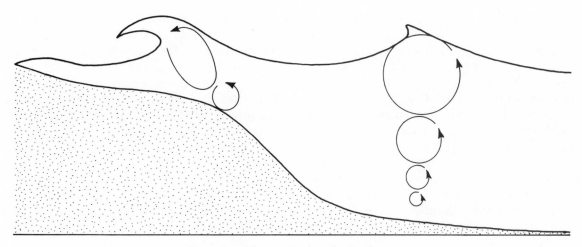

FIG. **10-1.** The mechanics of a breaker.

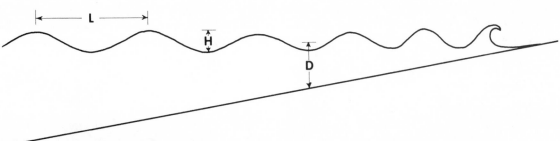

FIG. **10-2.** Properties of waves: L = Wavelength, H = wave height, D = water depth.

water, a circular wave pattern is generated behind the breakwater by diffraction. The diffracted waves intersect other incoming waves, resulting in an increased wave height.

The shape of a wave varies with water depth (FIG. 10-2). In deep water, a wave is symmetrical, with a smooth crest and trough. In shallow water, a wave is asymmetrical, with a peaked crest and a broad trough. Wavelength is measured from crest to crest and depends on the location and intensity of the storm at sea. The average lengths of storm waves varies from 300 to 800 feet. When waves move out of a storm area, the longer waves move ahead of the storm and form swells. On the open ocean, swells with wave lengths of 1000 feet are common, and they can reach a maximum of about 2500 feet in the Atlantic and about 3000 feet in the Pacific.

Wave height is measured from the top of the crest to the bottom of the trough. Although occasional storm waves of 30 to 50 feet high have been reported, they are not very common, and generally, wave heights are less than 20 feet. Exceptionally large ocean waves are rare. One such wave, reported in the Pacific by a U.S. Navy tanker in 1933, was over 100 feet high.

The *wave period* is the time it takes for a complete wave to pass a point. It is measured from one wave crest to the next. Periods for waves in the ocean vary from less than 0.1 second for small ripples to more than 24 hours. Waves with periods greater than 0.1 second but less than 5 minutes are called *gravity waves* and include the wind-driven waves that break against the coastline, most of which have periods between 5 and 20 seconds. Waves with periods between 5 minutes and 12 hours are *long waves*, which are often the result of storms.

A *seiche* is a wave caused by a sudden change in barometric pressure on a lake or large bay where the water sloshes back and forth. Seiches are common on Lake Michigan, and on occasions, they can be quite destructive (FIG. 10-3).

A seismic sea wave from an undersea earthquake or landslide usually has a period of 15 minutes or more and a wave length of up to several hundred miles. Tidal waves, with periods of from 12 to 24 hours, are set up in the oceans by the pull of gravity from the Sun and Moon. Other long waves are caused by seasonal differences in barometric pressure over different parts of the ocean, such as the Southern Oscillation discussed in Chapter 7.

Wave steepness is the ratio of wave length to wave height and is one of the most important aspects of waves. Storm waves with high steepness have short wavelengths and high wave heights, and are choppy. Swells have long wavelengths and low wave heights, and are smooth. Short waves, which are relatively steep, are particularly dangerous to small boats because the bow might be on a crest while the stern is in a trough.

Steep waves that accompany storms cause erosion of sea cliffs and sand dunes along the coast. Swells with low steepness generally result in the shoreward transport of sediment. Therefore, sediment that was carried offshore by storm waves is returned by swells in the interval between storms.

The wave motion changes as the wave moves from deep water toward the shore. The waves transport energy, but do not transport the water itself. As the wave crest approaches, an object floating on the surface first rises and moves forward with the crest, and after the crest has passed, the object then drops into the trough and moves backward. Thus, a floating object describes a circular path, with the diameter equal to the wave height and returns to its original position when the wave passes.

(Courtesy of NOAA)

FIG. **10-3.** Shoreline erosion on Lake Michigan.

The maximum sea state when waves reach their maximum height is usually achieved after 3 to 5 days of strong, steady winds from a storm that blows across the surface of the sea. The wave height is determined by the wind speed and duration. A fully developed sea can be reached in 24 hours with a wind speed of 30 knots, creating wave heights up to 20 feet. If, however, a sustained wind blew at 60 knots, a fully developed sea would have waves that average over 60 feet high.

The distance over which the wind blows on the surface of the ocean is called the *wave fetch*. It is dependent on the size of the storm and on the size of the body of water. In order for waves to reach a fully developed sea state, the fetch must be at least 200 miles for a wind of 20 knots, 500 miles for a wind of 40 knots, and 800 miles for a wind of 60 knots.

As the waves leave the storm area, they develop into swells that travel over great distances, sometimes halfway around the world, before they either die out or arrive at the coast. Waves with longer periods travel at faster speeds than waves with shorter periods. The speed is proportional to the square root of the wavelength. As the waves spread out from the storm area, they are sorted with the longer period waves out in front and the shorter period waves trailing behind. Therefore, as swells move across the ocean toward distant shores, the low, long-period waves are the first to arrive.

Waves that expand outward from a storm form rings similar to those formed by tossing a rock into a quiet pond. As the rings enlarge, the wave spreads out along a greater length and the circumference of the circle expands, causing the wave height to decrease as it moves away from the storm area. When the swells arrive at the coast, they form a uniform succession of waves, each with about the same period and height, which changes when the slower swells begin to arrive.

When the swells reach the coast, they form various types of breakers, depending on the wave steepness and bottom-slope conditions near the beach (FIG. 10-4). If the slope is relatively flat, less than 3 degrees, the wave will break and form a *spilling breaker*; that is, an over-steepened wave that starts to break at the crest and continues to break

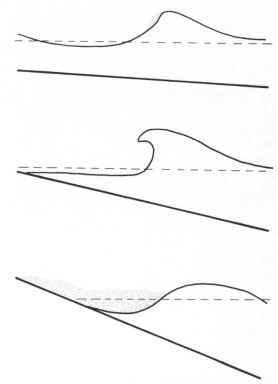

FIG. 10-4. Types of breakers: Spilling breaker (top), plunging breaker (middle), and surging breaker (bottom).

as the wave travels toward the beach. Spilling breakers are the most common type of breakers and also provide good waves for surfing.

A *plunging breaker* forms when the bottom slope is between 3 and 11 degrees, and the crest curls over, forming a tube of water. As the wave breaks, this tube moves toward the bottom beneath the wave and stirs up bottom sediments. Plunging waves are the most dramatic breakers and do the most damage because wave energy is concentrated where the wave breaks.

In a *collapsing breaker*, the bottom slope is relatively steep, between 11 and 15 degrees. The breaker is confined to the lower half of the wave. As the wave moves toward the coast, however, most of it is reflected off the beach.

A *surging breaker* develops on a steep bottom where the slope is greater than 15 degrees. The wave does not break, but surges up the beach face and is reflected off the beach, generating standing waves near the shore. Standing waves are important in the development of offshore bars, sand spits, beach cusps, and rip tides.

THE RISE AND FALL OF THE SEA

Tides are caused by the tug of gravity from the Moon and Sun on the ocean, and have a large influence on many aspects of life along the coast. The Moon revolves around the Earth in an elliptical orbit and exerts a stronger pull on the near side of the Earth than it does on the far side. The difference between the gravitational attraction on the two sides is about 13 percent, which tends to elongate the center of gravity of the Earth-Moon system. The inertia of the Earth as it orbits the center of gravity of the Earth-Moon system tends to stretch the planet out of its spherical shape.

The liquid oceans respond more to such forces than does the solid crust. Hence, the oceans flow into two tidal bulges with one facing toward the Moon and the other facing away from it, and between them the water is shallower, giving the oceans an overall egg-shaped appearance. The water in the middle of the ocean is actually only raised by about 2.5 feet at maximum high tide, but because of a sloshing-over effect and the configuration of the coastline, the tides on the coasts can be several times higher.

Since the Moon takes about 27.5 days to orbit the Earth, the tidal bulges take the same period as they follow the Moon around the planet. The daily rotation of the Earth on its axis causes each point on the surface of the Earth to go into and out of each tidal bulge once each day. As the Earth spins into and out of the two tidal bulges, the tides appear to rise and fall twice daily. The Moon also orbits the Earth in the same direction the Earth revolves. By the time a point on Earth has revolved halfway around, the tidal bulges have moved forward with the Moon, and the point must move a little farther each day in order to enter the bulge again. Therefore, there is slightly more than 12 hours between high tides; the actual period is 12 hours 25 minutes.

If there were no continents to impede the motion of the tides, all coasts would have two high tides and two low tides of nearly equal magnitude and duration each day. These are called *semidiurnal tides* and occur at places such as along the Atlantic coasts of North America and Europe.

Other places have different tidal patterns. The wave of the tide is reflected off and broken up by the continents and forms a complicated series of crests and troughs thousands of miles apart. Moreover, in some regions, the tides are coupled with the motion of large, nearby bodies of water. As a result, some places, such as along the coast of the Gulf of Mexico, have only one tide a day, which is called a *diurnal tide* and has a period of 24 hours 50 minutes.

Mixed tides are a combination of semidiurnal and diurnal tides, such as those which occur along the Pacific coast of North America. They display a diurnal inequality with a higher high tide, a lower high tide, a higher low tide, and a lower low tide each day. Some deep-draft ships on the West Coast must wait until the higher of the two high tides comes in before departing.

A few places, such as Tahiti, have virtually no tide. They lie on a *node*, which is a stationary point about which the standing wave of the tide oscillates.

The Sun also raises tides with diurnal and semidiurnal periods of 24 and 12 hours. Since the Sun is so much farther from the Earth than the Moon, its tides are only about half the magnitude of lunar tides. The *overall tidal amplitude*, which is the difference between the high-water level and the low-water level, depends on the relation of the solar tide to the lunar tide and therefore on the relative positions of the Sun, Moon, and Earth (FIG. 10-5).

The tidal amplitude is at its maximum twice a month, during the new moon and the full moon, when the three bodies are aligned in a nearly straight-line configuration known as *syzygy* (from the Greek word *syzygos*, meaning yoked together). This is the time of the *spring tides*, which have nothing to do with the season. The term comes from the Saxon word *springan*, meaning a rising or swelling of water.

Neap tides occur when the amplitude is at its minimum, during the first quarter and third quarter of the moon when the Sun, Moon, and Earth form

C-1. Breakers on the Pacific Ocean. (NOAA)

C-2. Ice floes in Antartica. (USAF)

C-3. (*Above*) Research vessel ATLANTIS II and submersible ALVIN. (WHOI)

C-4. (*Left*) Deep-sea vessel ALVIN underwater. (WHOI)

C-5. Giant crab feeding on giant tube worms on East Pacific Rise, Galapagos Islands. (WHOI)

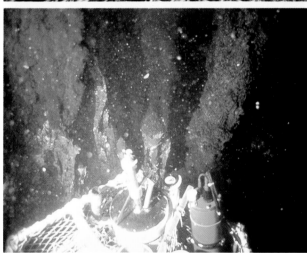

C-6. Black smokers on East Pacific Rise, Galapagos Islands. (WHOI)

C-7. Pillow lava on the ocean floor near Easter Island. (WHOI)

C-8. The research vessel KNORR near an iceberg. (WHOI)

C-9. Storm at sea. (USAF)

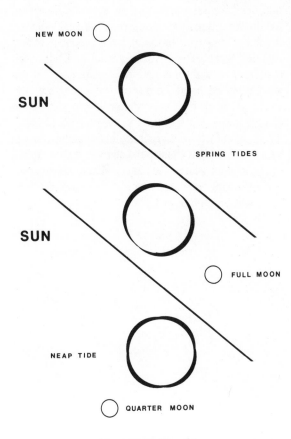

SUN

NEW MOON

SUN

SPRING TIDES

SUN

FULL MOON

NEAP TIDE

QUARTER MOON

FIG. 10-5. The tides.

a right angle and the solar and lunar tides oppose each other.

A tidal basin near the mouth of a river can resonate with the incoming tide. The effect of this natural oscillation is to make the water at one end of the basin high at the beginning, low in the middle, and high again at the end of the tidal period. The incoming tide sets the water in the basin oscillating, with water sloshing back and forth. The motion of the tide coming in toward the mouth of the river and the motion of the oscillation are synchronized. Thus, the oscillation reinforces the tide in the bay and makes the high tide higher and the low tide lower than it would be otherwise.

An excellent example of such a resonance is the Bay of Fundy in southeast Canada between New Brunswick and Nova Scotia. The bay is known for the world's highest tides, with a tidal range of more than 50 feet. When the tide is out, fishing boats not out to sea are left on the bottom of the harbor several stories below their wharfs. Cook Inlet in Alaska resonates so strongly that the normal 14-foot tide at its mouth is amplified more than 30 feet by the time it reaches Anchorage.

A special feature of this type of oscillation within a tidal basin is the *tidal bore* (FIG. 10-6). It is a solitary wave that carries a tide upstream, usually during a new or full moon. One of the largest tidal bores sweeps up the Amazon River and can be as high as 25 feet, as long as several miles across, and can reach 500 miles upstream. Although any body of water with high tides can generate a tidal bore, about half of the known tidal bores are associated with resonance in a tidal basin. The seaward ends of many rivers have tides.

At the mouth of the river the tides are symmetrical, and ebb and flood tide last about 6 hours each. *Ebb* and *flood* tides refer to the currents associated with the tides. Ebb currents flow out to sea, and flood currents flow into an inlet. Upstream, the tides become increasingly asymmetrical, with less time elapsing between low water and high water than between high water and low water. This is because the tide comes in quickly, but goes out gradually with the current. A tidal bore tends to exaggerate this asymmetry because the tide comes up the river very rapidly in a single wave. Therefore, the tides and their resonance with the oscillation in a tidal basin provide the energy for the tidal bore.

The incoming tide arrives in a tidal basin in the form of rapidly moving waves with long wavelengths. As the waves enter the basin, they are confined at both the sides and the bottom by the narrowing estuary. As a result of this funneling action, the height of the wave increases.

The increase in height is much like the development of breakers from waves moving onto a shelving beach. As the tide comes in and interacts with the bottom, refraction slows down the waves. Because the offshore waves are in deeper water, they move faster than the waves closer to shore, and as a result, the distance between successive wave crests is reduced closer inshore. Furthermore, because the crest of each wave is in deeper water than are the adjacent troughs, it begins to overtake the trough out in front. When this happens, the wave is no longer symmetrical, and the leading side becomes steeper and the trailing side becomes flatter. If the difference in depth between the two sides of a wave is great enough, the crest begins to break into the trough ahead and the tide wells up higher than the river level and forms a tidal bore.

As the tidal bore moves upstream, it must move faster than the current of the river; otherwise it would be swept downstream. Sometimes the speed of the river current and the speed of the tidal bore are the same, and the tidal bore appears to be stationary.

All the while, the shape and the dynamics of the tidal bore are not affected by the speed of the river. As the tidal bore moves up river, it might have a breaking crest, or it might be a smooth, gently rounded wave. The advancing front of a tidal bore is neither straight nor uniform. Some parts of the wave front lag behind the main wave and some advance ahead of it. The variation is a result of the change in the depth of the river from bank to bank. The parts of the wave that are in deeper water advance faster than those in shallower water, which tends to offer more resistance. Where there is local shallowing near the shore of the river, the tidal bore will be higher and possibly breaking.

(Courtesy of NOAA)

FIG. 10-6. Tidal bore at Moncton, New Brunswick, Canada.

TABLE 10-1. A Catalog of Tidal Bores.

COUNTRY	TIDAL BASIN	TIDAL BODY	KNOWN BORE LOCATION
Bangladesh	Ganges	Bay of Bengal	
Brazil	Amazon	Atlantic Ocean	
	Capim		Capim
	Canal Do Norte		
	Guama		
	Tocantins		
	Araguari		
Canada	Petitcodiac	Bay of Fundy	Moncton
	Salmon		Truro
China	Tsientang	East China Sea	Haining to Hangchow
England	Severn	Bristol Channel	Framilode to Gloucester
	Parrett		Bridgwater
	Wye		
	Mersey	Irish Sea	Liverpool to Warrington
	Dee		
	Trent	North Sea	Gunness to Gainsborough
France	Seine	English Channel	Caudebec
	Orne		
	Coueson	Gulf of St. Malo	
	Vilaine	Bay of Biscay	
	Loire		
	Gironde		Iles de Margaux
	Dordogne		La Caune to Brunne
	Garonne		Bordeaux to Cadillac
India	Narmada	Arabian Sea	
	Hooghly	Bay of Bengal	Hooghly point to Calcutta
Mexico	Colorado	Gulf of California	Colorado Delta
Pakistan	Indus	Arabian Sea	
Scotland	Solway Firth	Irish Sea	
	Forth		
U.S.	Turnagain Arm	Cook Inlet	Anchorage to Portage
	Knik Arm		

Eventually, friction between the water and the bottom, and the dissipation of energy into turbulence weakens the tidal bore. Its speed becomes less than that of the river, and the tidal bore is carried downstream and out to sea.

THE BIG PUSH

In addition to the pull of the Moon and Sun, the height of the tides are greatly affected by the weather. Storms produce the most significant discrepancies between predicted and actual heights of high water and low water at ports and seaside piers. The two major factors involved are pressure changes and strong winds. Low pressures, which raise the sea surface level, are characteristic of storms, which in turn are always accompanied by strong winds. Onshore winds, which blow toward the coast, push the water onto the shore. If they occur during high tide, sufficient water is piled up to cause severe flooding. Just the opposite effect occurs when very strong offshore winds, which blow toward the sea, lower the sea below the level of low tide sufficiently to ground vessels in port.

Most of the high waves and beach erosion occur during coastal storms. Thunderstorms and squalls are two of the most violent storms along the coast. They are most frequent in the midlatitudes and produce gusty winds, hail, lightning, and a rapid buildup of seas. The life cycle of a single thunderstorm cell is usually less than half an hour. When the cell dies, a new one often develops to take its place.

Frontal thunderstorms are formed at the leading edge of a cold front and move in a line abreast about 25 miles per hour. A squall line often precedes a cold front as far out as 200 miles. It has a distinctive dark gray, cigar-shaped cloud that appears to roll across the sky from one end of the horizon to the other. Squall lines move about 25 miles per hour, and the winds in a squall can reach 60 miles per hour, but they are generally short-lived, and are usually over in less than 15 minutes. Within minutes after a squall arrives, waves several feet high are produced, but since the winds in a squall do not last for long, the waves die down almost as rapidly as they build up.

Tropical cyclones, or hurricanes, produce the most dramatic storm surges (FIG. 10-7). Hurricane-force winds caused by the rotation and forward motion of the storm reach 100 miles per hour or more

(Courtesy of NOAA)

FIG. 10-7. Storm surge from a hurricane.

and push water out in front of the storm. The low pressure in the eye of the hurricane draws up the water into a mound several feet high. As the hurricane moves across the ocean, it can set up a resonance with the swells it generates if the hurricane speed matches the speed of the waves. This adds to the height of the swells, and swells over 60 feet have been reported in hurricanes. When the hurricane approaches the coast, the effect of the water piled up by the wind, the mounding of water by the low pressure, and the generation of swells and the possible resonance of swell waves can make a most deadly combination when they are superimposed on the regular cycle of tides. They can result in massive flooding, devastation of property (FIG. 10-8), and the loss of a great many lives.

The result of all this lapping of seawater against the shore during a serious storm is coastal erosion (FIG. 10-9). The cliffs and dunes that mark the coastline are eroded, causing the shoreline to retreat. Between 1888 and 1958, the coastline on Cape Cod, Massachusetts, from Nauset Spit to Highland Light retreated at an average rate of over 3 feet per year. The soft cliffs of the Suffolk Coast of England on the North Sea are eroding at an average rate of 10 to 15 feet per year. At the town of Lowestoft, a 40-foot-high cliff of unconsolidated rock was eroded back 40 feet, and where the cliff was only about 6 feet high, it was eroded back 90 feet, all by a single storm.

Beach erosion is very difficult to predict and almost impossible to stop. It is so variable because it depends on the strength of the dunes or cliffs, the intensity and frequency of the storms, and the exposure of the coast. Most attempts to prevent beach erosion are doomed from the very start because the waves constantly batter and erode any man-made defenses.

(Courtesy of NOAA)

FIG. 10-8. Storm surge damage to homes at Virginia Beach, Virginia.

The beach's rate of retreat varies with the shape of the shore and the prevailing wind and tides. More than half of the 72-mile-long south shore of Long Island, New York, is considered a high-risk zone for development. Some locations there are being reclaimed by the sea at a rate of 6 feet per year. The barrier island, running from Cape Henry, Virginia, to Cape Hatteras, North Carolina, has narrowed from both the seaward and landward sides. The rest of the North Carolina coast is moving back at 3 to 6 feet annually, and most of East Texas is vanishing even faster. In California, homes are falling into the sea, causing $100 million in property damage each year.

Some 80 to 90 percent of America's once-sandy beaches are sinking beneath the waves. Most of the problem stems from the methods engineers use to try to stabilize the beaches. Jetties cut off the natural supply of sand to beaches, and seawalls actually increase erosion by bouncing waves back without absorbing very much of their energy. The rebounding waves carry sand out to sea, undermining the beach and destroying the very shorefront the seawall was built to protect.

THE GREAT WAVE

Another kind of wave is produced by undersea and near-shore earthquakes, landslides, and volcanic eruptions. It is called a *seismic sea wave*, or *tsunami*, a Japanese word meaning "tidal wave." Actually, the waves really have nothing to do with the tides.

(Courtesy of U.S. Corps of Engineers)

FIG. 10-9. Beach wave erosion at Grand Island, Louisiana, from hurricanes Danny and Elana.

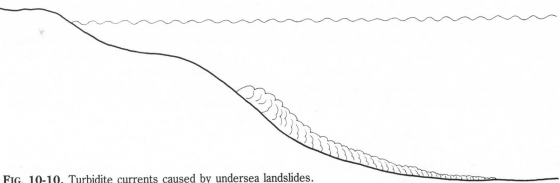

FIG. 10-10. Turbidite currents caused by undersea landslides.

The most destructive tsunamis are caused by the vertical displacement of the ocean floor. The energy of an undersea earthquake is transformed into wave energy proportional to the intensity of the earthquake. In the open ocean, the wave crests can be up to 300 miles long and are usually less than 3 feet high. The wavelength is between 60 and 120 miles, giving a tsunami a very gentle slope and allowing it to pass by ships practically unnoticed.

Tsunamis travel at speeds between 300 and 600 miles per hour. When a tsunami touches bottom in a harbor or narrow inlet, its speed diminishes rapidly to about 100 miles per hour. This sudden breaking action causes the water to pile up, and the wave height is magnified tremendously. Tsunamis have been known to grow into a wall of water up to 200 feet high, although most are only a few tens of feet high. The destructive force of the wave is immense, and the damage it causes as it crashes to shore is considerable. Large buildings are crushed with ease, and sizable ships are tossed up and carried well inland as though they were toys.

Coastal and submarine landslides also can cause tsunamis. Coastal landslides are produced when a sea cliff is undercut by wave action and falls into the ocean. Excessive rainfall along the coast also can lubricate sediments, allowing large blocks to slide into the sea.

Submarine landslides move down steep continental slopes and have been known to bury submarine telephone cables under thick layers of sediment. Submarine landslides also might be responsible for carving out deep submarine canyons. The landslides consist of sediment-laden water. Be-cause they are heavier than the surrounding water, they move along the ocean floor and are capable of eroding the soft bottom material. These muddy waters are called *turbidity currents*, and they can move down the gentlest slopes and transport fairly large rocks (FIG. 10-10). Turbidity currents also can be initiated by river discharge, coastal storms, or other currents. They play an important role in building up the continental slopes, as well as the smooth ocean bottoms at the foot of the continental slope.

Explosive volcanic eruptions associated with the birth or death of a volcanic island also can set up large tsunamis. The 1480 B.C. eruption of Thera set in motion a tsunami that was responsible for damaging, in part, Minoan settlements in Crete and elsewhere in the Mediterranean, and eventually led to the fall of the Minoan Empire. In the 1883 eruption of Krakatau, a giant tsunami was generated when the volcanic crater collapsed, causing millions of tons of seawater to rush into the hot magma chamber. Like a gigantic pressure cooker, the volcano blew its lid, producing the greatest explosion in modern times and destroying a large portion of the island. Three great tsunamis, up to 100 feet high, ripped through the coasts of Java and Sumatra, destroying coastal villages and taking the lives of 36,000 people. By far, the most tsunami-prone area in the world is the rim of the Pacific Ocean, which has the most earthquakes as well as the most volcanoes.

Up to about 40 years ago, earthquakes on the ocean floor went largely undetected, and the only warning people had of a tsunami was a rapid withdrawal of water away from the shore. Residents of coastal areas frequently stricken by tsunamis have learned to heed this warning and head for higher ground. During the great Lisbon, Portugal, earthquake of 1755, large quantities of fish were left high and dry when the sea retreated on the island of Madeira in the Azores. The villagers, unaware of any danger, went out to collect this unexpected bounty, only to lose their lives when the tsunami struck without warning.

A few minutes after the sea retreats, there is a tremendous surge of water, capable of extending hundreds of feet inland. There is often a succession of surges, each followed by a rapid retreat of water back to the sea. On coasts and islands where the seafloor rises gradually or where there are barrier reefs, much of the energy of the tsunami is spent before it ever reaches the shore. On volcanic islands that are surrounded by very deep water or where deep submarine trenches lie immediately outside harbors, an oncoming tsunami can build to prodigious heights.

Destructive tsunamis from large-magnitude earthquakes can travel all the way across the Pacific. The great Chilean earthquake of 1960 sent out a tsunami that struck Hilo, Hawaii, over 5000 miles away, with up to 35-foot waves, causing over $20 million in property damage and killing 61 people. The tsunami then traveled another 5000 miles to Japan and inflicted considerable destruction on the coastal

(Courtesy of NOAA)

FIG. 10-11. Tsunami damage in Seward, Alaska from the March 27, 1964, Alaskan earthquake.

villages of Honshu and Okinawa, leaving 180 people dead or missing. In the Philippines, 20 people were killed, and coastal areas of the United States and New Zealand were damaged. For several days afterwards, tidal gauges in Hilo could still detect the waves as they bounced around the Pacific basin.

The tsunami from the 1964 Alaskan earthquake destroyed coastal villages around the Gulf of Alaska, killing 107 people (FIG. 10-11). The tsunami damage extended along much of the west coast of North America, causing more than $100 million in damages. Despite plenty of warning, 11 people died in Crescent City, California, from a 20-foot wave.

A tsunami originating in Alaska can reach Hawaii in 6 hours, Japan in 9 hours, and the Philippines in 14 hours. A tsunami originating off the Chilean coast can reach Hawaii in 15 hours and Japan in 22 hours—time enough to take the necessary safety precautions that might make the difference between loss of life and property.

11

Treasures from the Seabed

Throughout geologic history, in many parts of the world a piece of ocean crust was scraped off an oceanic plate as it plunged under a continental plate and was plastered against the leading edge of the continent. These slices of ocean crust are called *ophiolites*, from the Greek word *ophis*, meaning "snake" or "serpent." The associated igneous rock is known as *serpentinite*, so-named because of its mottled green appearance.

When former volcanically active regions of the ocean crust are uplifted onto the continents, ophiolites provide rich metal ore deposits that are mined throughout the world. More than 3000 years ago, the Troodos ophiolite on Cyprus was mined extensively by the Greeks for its copper and tin. The ore provided some of the first bronze for the earliest Greek sculptures. In essence, the Greeks were mining the ocean floor that had been conveniently brought up to the surface.

HOT-WATER RICHES

Hydrothermal ore deposits are associated with volcanically active zones on the ocean floor (FIG. 11-1) such as spreading ridges, where new oceanic crust is formed, and island arcs, where the oceanic crust is destroyed in deep-sea trenches. These deposits have been found on young seafloor along active spreading centers of the major oceans, as well as regions that are rifting apart and forming new oceans such as the Red Sea, the Afar Rift, and the Gulf of Aden. Also, deep-sea drilling has uncovered identical deposits in older ocean floor far from modern spreading centers. This indicates that the process responsible for the creation of metal deposits has operated throughout the history of the major oceans, and is also further evidence for seafloor spreading.

The deposits were formed by the precipitation of metals from hot-water solutions that were rich in silica and metal and were discharged through the seafloor by hydrothermal springs (FIG. 11-2). Silica and other minerals build prodigious chimneys as tall as 30 feet, from which turbulent black clouds of fluid billow out. These chimneys are called black smokers. Metal-rich particles precipitated from the smoke eventually might fill a depression and form an ore body.

130

FIG. 11-1. Location of ore deposits originally formed by seafloor hot springs.

Seawater seeps into cracks in the basaltic rock on the ocean floor, is heated by hot rocks near a magma chamber a few miles down, and water with a temperature several hundred degrees Fahrenheit, rises to the surface by convection. Along the way, the hot water dissolves silica and metals from the surrounding rock. When the hot metal-rich solution emerges from a spring into cold, oxygen-rich seawater, metals such as iron and manganese are oxidized and deposited, along with silica. Some deposits from the Mid-Atlantic Ridge contain as much as 35 percent manganese, an important metal used in making steel alloys.

This type of hydrothermal deposit is generally poor in copper, nickel, cobalt, lead, and zinc because these elements tend to stay in solution longer than iron and manganese. Under oxygen-free conditions, such as those in stagnant pools of brine, copper and zinc can be concentrated along with iron and manganese. These deposits are found in the Red Sea where the concentrations of copper and zinc reaches a few percent, making them economically attractive.

The entire volume of the oceans undergoes hydrothermal circulation through the crust at spreading ridges every 10 million years. The annual flow is comparable to the annual flow of the Amazon River. This circulation accounts for the unique chemistry of seawater and for the efficient thermal and chemical exchanges between the oceanic crust and the ocean. The magnitude of some of these chemical exchanges is comparable to the input of elements into the oceans by rivers, carrying materials weathered from the continents.

The elements that are contributed to hydrothermal systems are ultimately derived from the mantle at depths of 20 to 30 miles. Magma upwelling from the mantle penetrates the ocean crust and provides new crustal material at the spreading centers. As seawater penetrates below the seafloor in the vicinity of the active zone, it circulates within the zone of young, highly fractured crust and is heated. The ocean crust is fractured because it contracts as it cools from its original molten state and because of the compressional forces, separating the two oceanic plates at the spreading centers. The heated water leaches a number of elements from the basalt and these are carried in solution up to the surface by convection and discharged through the seafloor. In addition, metal-rich fluids, called *juvenile water*, derived directly from the magma and volatile elements derived from the mantle are transported with the hydrothermal waters to the surface.

As the oceanic crust moves away from the spreading centers, it eventually reaches the margin of the ocean basin, where it is either subducted into the mantle or collides head-on with another lithologic plate and raises mountains. In the course of these events, fragments of oceanic crust might be uplifted and exposed on land. These fragments, or ophiolites, have been identified in various parts of the world.

Ophiolites consist of an upper layer of marine sediments, a layer of pillow lava which are basalts that have erupted under water, and a layer of dark, dense ultramafic rocks that are thought to be part of the upper mantle. The metal ore deposits (FIG. 11-3) are at the base of the sedimentary layer just above the area where it makes contact with the basalt. Examples include the 100-million-year-old ophiolite complexes exposed on the Apennines of northern Italy, the northern margins of the Himalayas in southern Tibet, the Ural Mountains in Russia, the eastern Mediterranean, including Cyprus, the Afar Desert of northeastern Africa, the Andes of South America, and islands of the western Pacific such as the Philippines, uppermost Newfoundland, and Point Sol along the Big Sur coast of central California.

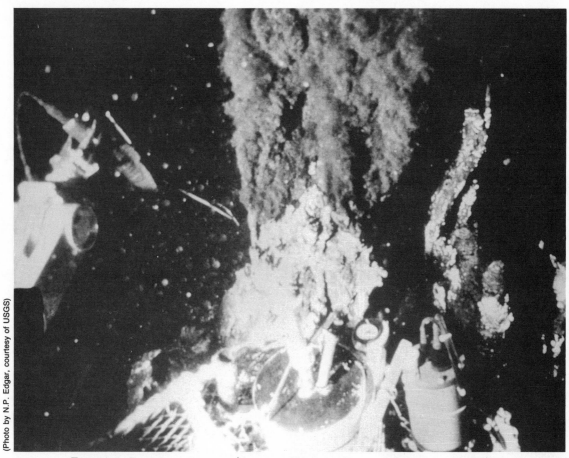

(Photo by N.P. Edgar, courtesy of USGS)

FIG. 11-2. Hydrothermal vent pouring out sulfide-laden hot water onto the ocean floor.

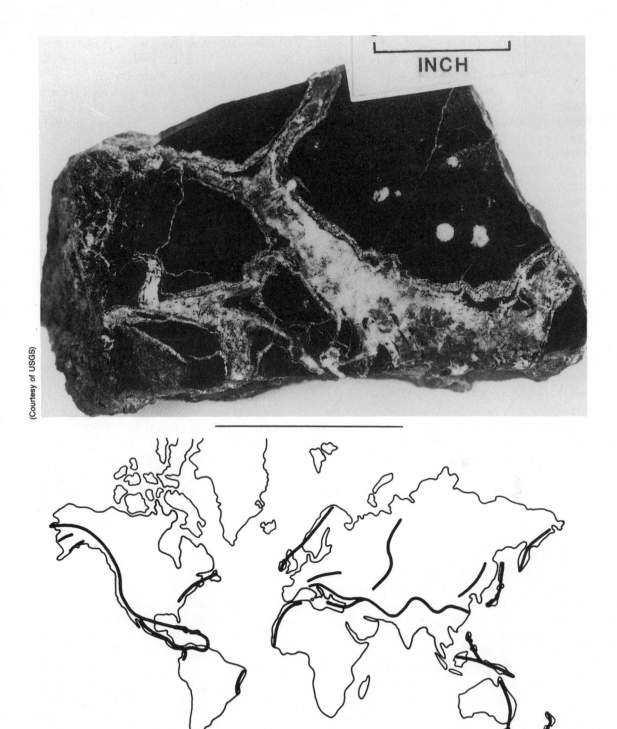

(Courtesy of USGS)

FIG. 11-3. Stockwork sulfide mineralization found in ophiolites (top), and worldwide distribution of ophiolites (bottom).

Another type of metal ore deposits formed at oceanic spreading centers are called *massive sulfides*. The deposits contain sulfides of iron, copper, lead, and zinc. They occur in most ophiolite complexes, such as the Apennine ophiolites, which were first exploited by the Romans. They are mined extensively in other parts of the world for their rich ores of copper, lead and zinc.

Magma contains sulfur. In the course of the circulation of seawater below the ocean floor, the waters acquire sulfate ions and become strongly acidic. This favors the combination of sulfur with certain metals leached from the basalt and extracted from the hydrothermal solution to form insoluble sulfide minerals. Sulfide metals are deposited from hydrothermal systems on the ocean floor to form large mounds (FIG. 11-4), or are deposited as disseminated inclusions in the rock below the seafloor called *veins*. Some of the world's most important deposits of copper, lead, zinc, chromium, nickel, and platinum, which are critical to modern industry, were originally formed several miles below the seafloor and were later upthrusted onto dry land.

Sulfide ores on the seafloor were first charted in the eastern Pacific in 1978 by the French research submersible *Cyana*, which discovered mounds of a porous gray and brown material up to 30 feet high in water over 1.5 miles deep. The massive sulfide deposits contained abundant amounts of iron, copper, and zinc.

Scientists on the research vessel *Sonne* found a sulfide ore field nearly 2000 miles long on the floor of the East Pacific. The sediments contained as much as 40 percent zinc. Other metals also were present in the deposits, some of which were in greater concentrations than their landward counterparts.

Following the discoveries, Germany and France set up a joint project called Geometep to investigate the ores. Despite the value of these finds, however, commercial exploitation of deep-sea ores might have to wait until the next century because of the present depressed prices for these commodities. Also, startup costs for planned operations have risen dramatically, making such mining ventures economically prohibitive.

Hot-water brines are the result of the opening of a new ocean basin and are associated with a slow spreading center such as the one that bisects the Red Sea (FIG. 11-5). There, hot metal-rich brines fill a number of basins along the spreading zone. The cold, dense seawater percolating down through volcanic rocks becomes unusually salty because it passes through thick beds of rock salt (sodium chloride) buried in the crust.

The salt beds formed under dry climatic conditions when evaporation exceeded the inflow of seawater that was nearly cut off by the surrounding landmass. When the salinity increased to the saturation point, salt crystals precipitated out of solution, settled on the ocean floor, and accumulated in thick beds.

The high salinity of hot circulating solutions through these salt beds enhances their ability to transport dissolved metals by forming complexes with the chlorine from the salt. When they are discharged from the floors of the basins, the heated solutions collect as hot brines. Metals precipitate from the hot brines and settle in basins, where they are trapped and form layered deposits of metalliferous sediments in places up to 6 miles thick.

In the 1970s, research vessels found valuable sediments on the bed of the Red Sea more than 7000 feet below sea level. The largest of these deposits is in a 3.5-mile-wide area known as the Atlantis II Deep, named for the research vessel which discovered it, and lies between Sudan and Saudi Arabia due west of Mecca. The rich bottom ooze was estimated to contain some 2 million tons of zinc, 400,000 tons of copper, 9000 tons of silver, and 80 tons of gold.

In 1976, the Saudi-Sudanese Red Sea Commission was established to exploit this mineral wealth with the West Germany mining firm Preussage as the operator. The idea was to use a large vacuum cleaner to sweep the ocean bottom. The vacuum uses a vibrating screen and jets of water to loosen the sediment, which is then brought to the surface by the suction from a powerful vacuum pump. In 1979, a mining feasibility study was conducted and a full-scale pilot mining operations will begin when the world market becomes more favorable.

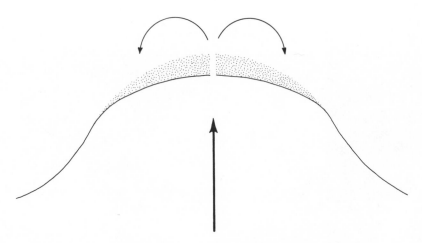

FIG. 11-4. A weathered sulfide mound on the Juan de Fuca Ridge (top),
and formation of a massive sulfide deposit   (bottom).

THE COBBLED SEAFLOOR

In the 1870s, the research vessel H.M.S. *Challenger* was dredging the ocean bottom in the Pacific when it brought up something that resembled dense lumps of coal. Scientists mistook them for fossils or meteorites and they ended up as geological oddities in the British Museum. Eighty years later, scientists realized the value of the dark, potato-size lumps. They were found to contain important quantities of valuable metals, including manganese, copper, nickel, cobalt, and zinc. The world's largest reserve of these metals was discovered by the research vessel *Glomar Challenger* on the bottom of the North Pacific about 16,000 feet below the

(Courtesy of USGS)

Fig. 11-5. The Red Sea viewed from Gemini spacecraft.

surface. Fields thousands of miles long contained manganese nodules (FIG. 11-6) estimated to contain 10 billion tons of rich ore with 10 million new tons formed every year.

Manganese nodules are hydrogenous (from Greek, meaning "generated from water") deposits formed by the slow accumulation on the ocean floor of metallic elements extracted directly from seawater. The water in the oceans contains minute quantities of metals such as iron and manganese in solutions with concentrations of less than one part per million by weight. Most metallic elements have a very limited solubility in an alkaline and oxygen-rich environment such as seawater, so they are easily extracted from the water. The extraction occurs when dissolved iron and manganese are oxidized by the oxygen in the seawater, forming insoluble oxides and hydroxides of iron and manganese. The metals are then deposited on the ocean floor as tiny particles or as films or crusts on any solid material on the seafloor. Certain metals also can be extracted from seawater by living organisms. After the organisms die, their remains collect on the ocean floor, where the metals are released and become incorporated in the bottom sediments.

The growth rate of hydrogenous deposits is exceedingly slow, sometimes less than 0.1 inch per million years. Most of the seafloor concretions such as manganese nodules are particularly well developed in deep, quiet waters far removed from volcanic ridges and spreading centers. They exist in regions of the basins that receive a minimal inflow of sediments from the land, which tends to bury them. Such areas include abyssal plains far from the continents and elevated areas on the ocean floor such as seamounts and isolated shallow banks. Elsewhere, the steady rain of clay and other mineral particles prevents the metals from growing into concentrated deposits.

The manganese nodules tend to grow around a solid nucleus, or seed, such as a grain of sand, a piece of shell, or a shark's tooth. The seed acts as a catalyst, which allows the metals to accrete to it like the growth of an oyster pearl. Concentric layers are added until the nodules reach the size of cobblestones, giving the ocean floor a cobbled appearance.

Every ton of manganese nodules contains about 600 pounds of manganese, 29 pounds of nickel, 26 pounds of copper, and 7 pounds of cobalt. If these

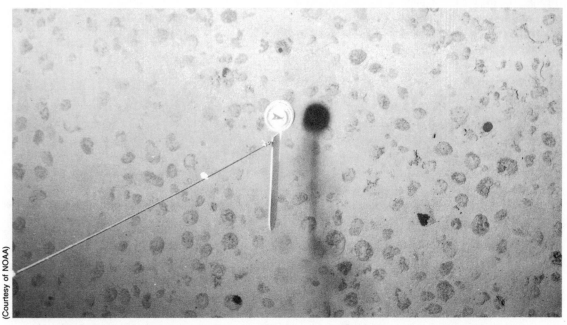

(Courtesy of NOAA)

FIG. 11-6. Manganese nodules on the ocean floor at a depth of 3 miles in the equatorial Pacific belt.

deposits were on land, there would be a mad rush to get at them. Because they are as deep as 4 miles below the sea, however, the nodules are not nearly so appealing. It requires the sifting of 100 square yards of ooze in order to extract a single ton of nodules.

One method of extraction calls for the use of a dredge to scoop up the nodules. Another approach is similar to the one used on the Red Sea where a gigantic vacuum cleaner is employed to suck up the nodules. A more exotic scheme envisions the use of television-guided robots to rake up the nodules, which are crushed and pumped to the surface in a slurry.

With all this mucking around down below, some environmentalists are concerned that clouds of sediments raised by the mining activities might disrupt the delicate ecology of the seafloor. Also, law-of-the-sea disputes over the ownership of midocean ore deposits have reduced the interest of western industrial nations, leaving the future of undersea mining and refining of manganese nodules in the hands of the Asian countries, including Japan, China, South Korea, and India, which need them in order to free themselves of dependence on foreign raw materials. Because of their lack of mineral resources, New Zealanders want to mine calcium phosphate nodules discovered east of New Zealand and grind them up to make fertilizer.

SALT-WATER MINERALS

Seawater contains about 3.5 percent dissolved minerals, mostly common salt. Salt has been mined extensively throughout the world since ancient times from *evaporite deposits*, which once were shallow, stagnant pools of seawater called *brines* that eventually evaporated. The evaporation occurs in shallow, slowly sinking basins, partially dammed by sandbars. During a storm, seawater flows over the sandbars and replenishes the basin.

Salt also accumulates in thick beds in deep basins (FIG. 11-7) that have been cut off from the general circulation of the ocean. As seawater evaporates, the concentration of salt increases to the saturation point, the salt precipitates out of solution, and accumulates on the seafloor. Several

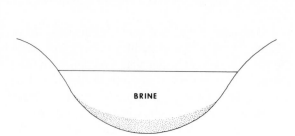

FIG. 11-7. Formation of salt deposits in a deep-ocean basin.

thousand feet of seawater must evaporate in order to produce 100 feet of salt; however, many salt deposits are much thicker than this. Several other minerals also are deposited, including gypsum (used in wall board), nitrates and phosphates (used in fertilizers and explosives), potassium (used in a wide variety of chemicals such as fertilizers), and important chemical elements such as bromine, chlorine, and iodine.

Evaporite deposits generally form under arid conditions between 30 degrees north and south of the equator. However, extensive salt deposits are not being formed at the present time. The fact that evaporites exist as far north as the arctic regions indicate that either these regions were at one time closer to the equator or the climate was considerably hotter in the geologic past than it is today. Also, evaporite deposits in the interiors of continents, such as the potassium deposits at Carlsbad, New Mexico, indicate that these areas were at one time inundated with seas.

The salts precipitate out of solution in stages. The first mineral to precipitate is calcite, which is closely followed by dolomite, although only small amounts of dolomite and limestone are formed in this manner. After about two-thirds of the water is evaporated, gypsum is precipitated. When nine-tenths of the water is removed, halite, or common salt, forms. During the last stages of precipitation, potassium and magnesium salts form.

At present, the Mediterranean Sea is an almost closed body of water, and the evaporation rate is

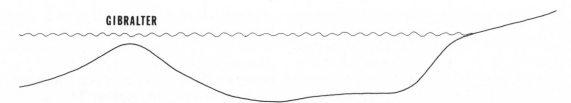

FIG. 11-8. Profile of the Mediterranean Basin looking north.

very high: nearly 5 feet of the water's surface evaporates every year. This produces water with a high salt content, which is heavy and sinks to the bottom and will eventually fill the basin. The inflow from rivers into the Mediterranean cannot compensate for the evaporation and the outflow of water at the Gibralter shelf (FIG. 11-8).

Six million years ago, the entire Mediterranean Basin was cut off from the Atlantic when Gibralter was uplifted, forming a dam across the strait. The whole sea evaporated, leaving a gigantic hole over 1 mile deep. At the bottom, salt deposits formed when the salt from the entire water column precipitated and was deposited on the seafloor. Afterwards, Gibralter subsided, the dam was broken, and the sea slowly refilled the basin. Subsequent sedimentation has buried trillions upon trillions of tons of salt beneath younger sediments.

The Gulf of Mexico was once completely dry 140 million years ago. The North Sea and the Red Sea also show signs of evaporation and are floored with thick layers of salt.

BLACK GOLD OFFSHORE

Among all the wealth waiting under the waves, only oil and natural-gas fields in shallow coastal waters have become profitable. The first interest in offshore drilling for petroleum came in the mid-1960s, and drilling stepped up considerably a decade later, following the Arab oil embargo of 1973. New important finds such as Prudhoe Bay on Alaska's North Slope (FIG. 11-9) and on the North Sea off Great Britain came out of intensive exploration for new reserves of offshore oil. The desire for energy independence at any price set western nations looking for oil in the deep oceans, where the difficulties encountered such as storms at sea and

the loss of personnel and equipment could not justify the few finds.

About 20 percent of the world's oil and about 5 percent of the natural-gas production is offshore. Plans are being formed to put drilling equipment and workrooms together on the seafloor where they are not affected by storms, making some deep-sea oil and gas fields available for the first time. Projections indicate that, in the future, perhaps twice as much oil will be pumped from the seas as from the land. Unfortunately, a lot of offshore oil leaks into the oceans—as much as 2 million tons each year—could prove to be an ecological disaster as production increases.

Large reservoirs of oil and natural gas are dependent on the existence of certain geological conditions, such as a sedimentary source for the oil, a porous rock to serve as a reservoir, and some kind of confining structure. The source material is generally organic carbon in fine-grained, carbon-rich sediments. Porous and permeable sedimentary rock such as sandstone often provides the reservoir. Diverse geologic structures that result from folding or faulting of sedimentary layers can serve as dams to pool the oil.

Most of the organic material capable of forming oil comes from microscopic plants and animals that originated primarily in surface waters and were concentrated in fine particulate matter. In order for organic material to be transformed into oil, at least one of two conditions must be met. Either the rate of accumulation must be so high or the oxygen in the bottom water must be so low that the material is not oxidized before it is buried. Oxidation causes decay, which destroys organic material; therefore, areas where there are high rates of accumulation of sediments rich in organic material are the most favorable sites for the formation of oil-bearing rock.

After deep burial in a sedimentary basin, the organic material is baked in a giant pressure cooker, which chemically alters it. In essence, the organic material is "cracked" into hydrocarbons by the heat from hydrothermal activity. If the hydrocarbons are overcooked, gas will result.

Oil is often associated with thick beds of salt. Because salt is lighter than the overlying sediments, it rises toward the surface, creating salt domes that help trap the oil. Hydrocarbon volatiles, along with seawater locked up in the sediments, migrate upwards through permeable rock layers and accumulate in traps formed by sedimentary structures that provide a barrier to further migration. In the absence of such *caprock*, the volatiles will simply continue to rise to the surface and escape through the ocean floor. It takes from several tens of millions to a few hundred million years for this process to take place.

The length of time mainly depends on the temperature and pressure conditions within the sedimentary basin. Plate tectonics plays an important role in determining whether these conditions are met, and greatly aids the geologist in his exploration activities.

When an oil company explores the ocean for oil and gas, it first searches for sedimentary structures conducive to the formation of oil traps. Exploration starts with a seismic survey, in which waves similar to sound waves are generated by explosives set off on the ocean floor or by air guns and received with hydrophones towed behind a ship (FIG. 11-10). The seismic waves are reflected and refracted off various sedimentary layers, giving the exploration geologist a sort of geological CAT scan of the ocean crust.

Once a suitable site has been located, a drill rig is brought in, which either stands on the ocean floor

(Courtesy of U.S. Maritime Administration)

FIG. 11-9. The Valdez, Alaska, terminal of the trans-Alaska pipeline for North Slope oil.

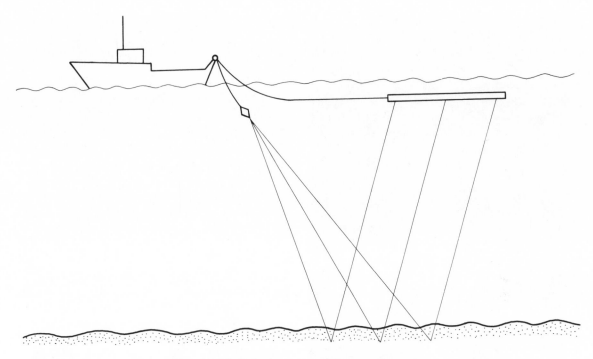

FIG. 11-10. Seismic survey of the ocean's crust.

if the water is shallow or free-floats while anchored to the bottom if the water is deep (FIG. 11-11). As the hole is being drilled through the bottom sediments, it is lined with steel casing to prevent cave-ins and act as a conduit for the oil. A blowout preventer is placed on the ocean floor on top of the casing to prevent the oil from gushing out once the drill bit penetrates the caprock. If the well is successful, a pumpdown test is conducted to determine its production, and more wells are drilled to fully develop the field.

In 1981, the Department of the Interior estimated that 27 billion barrels of oil and 163 trillion cubic feet of natural gas remain to be discovered in offshore deposits that are large enough to be commercially exploited around the United States. Estimates of undiscovered oil resources are by their very nature uncertain, and are based largely on geological data and a lot of informed guesswork.

In 1985, the department reduced its estimates of oil reserves in offshore fields by about half: 12.2 billion barrels of oil and 90 trillion cubic feet of gas. The new figures reflect the fact that oil companies

came up with nearly 100 dry wells over the previous 4 years of exploratory drilling in highly promising areas of the Atlantic and off the coast of Alaska. These regions have long figured prominently in projections of American oil supplies, and without that oil, the country could become dangerously dependent on foreign sources.

LAW AND ORDER ON THE SEA

Throughout modern history, freedom of the seas could only be enjoyed by those nations powerful enough to enforce it. Just after World War II, which was basically fought by the United States in order to preserve its freedom of the seas, the United States started the expansion of national claims to the ocean and its resources. The Truman Proclamations on the Continental Shelf and the Extended Fisheries Zone of 1945 set nations into a scramble to carve up the world's oceans, just as Africa had been carved up a century earlier. Meanwhile, a procession of new sovereign nations came out of the colonial empires assembled by former marine powers of Europe. The poorer nations of the world, which make up the

FIG. 11-11. A semisubmersible drilling rig.

majority, were bound and determined to have their say concerning the law of the sea even though a number of them were land-locked countries.

On December 6, 1982, 119 nations signed the United Nations Convention on the Law of the Sea, a kind of a constitution for the ocean. The Convention put 40 percent of the ocean and its bottom adjacent to the coasts of the continents and islands under the management of the states in possession of those coasts. The other 60 percent of the surface and the water below was reserved for the traditional freedom of the seas, and the wealth of the ocean floor, which makes up 42 percent of the Earth's surface, was deeded to the Common Heritage of Mankind. The Convention placed that heritage under the management of an International Seabed Authority, which has the capacity to generate income, the power of taxation, and a sort of eminent domain over ocean-exploiting technology.

The Convention provided a comprehensive global framework for the protection of the marine environment, a new regime for marine scientific research, and a comprehensive system for settling disputes. It ensured freedom of navigation and free passage through straits used for international navi-

FIG. 11-12. The world's economic zones.

gation, a right that cannot be suspended under any circumstances. In essence, the Convention provided a new order that is more responsive to the real needs of the world than the old order, which was disintegrating in hunger and turmoil.

Coastal states were accorded a 12-mile limit of territorial sea, as well as a 12-mile contiguous zone. Beyond these limits, they were accorded a 200-mile economic zone (FIG. 11-12) that includes fishing rights and rights over all resources. In cases where the continental shelf extends beyond the 200-mile limit, the economic zone with respect to resources on the shelf is extended to 350 miles.

The economic zone provides an economic incentive for managing the marine environment. Unfortunately, neither fish nor pollution respect national boundaries. The economic zone concept has been called the greatest territorial grab in history, giving coastal states unfair advantage, which increases inequality among nations. The discovery of a significant resource anywhere in the world's ocean will invite a claim from the nearest coastal or island state, even though it lies beyond the limits of national jurisdiction.

The freedom of the seas to do scientific research is also constrained by the expansion of national jurisdictions into the oceans. Citizens of other nations must apply for consent from a coastal state to do research in waters that were once open to all. Opposition to such a scientific project by a coastal state might undermine the cooperative atmosphere among nations that the Law of the Sea was trying to foster.

12

Energy
from the Ocean

ONE of the keys to industrialization is the capacity to generate power. As fossil fuels become scarce, however, there is a need to find alternative sources of power. For nearly two millenia after the waterwheel was first introduced by the Romans, man has used the power of falling water to drive his machines. Presently, falling water produces about one-fourth of the world's electrical power. Hydroelectric energy is now and will continue to be an important component of energy production.

The power of rivers and streams had its beginning in the ocean. When the Sun evaporates seawater, it imparts potential energy to each molecule of water vapor by raising it from sea level to several thousand feet. As water vapor condenses into rain, it gives up some of this potential energy as heat. When a raindrop strikes the ground, a certain amount of energy is lost by the impact, which causes erosion. As a river flows toward the ocean, most of the remaining energy is lost through friction. By the time the water reaches the sea again, it has lost practically all of its original energy.

The Sun also drives the ocean's waves and currents, which can be harnessed to provide electricity. The solar and lunar tides likewise can be used to generate power. Wind is also generated by the Sun and sea, and windmills are an important source of energy. Sometime in the not too distant future, fusion energy will be produced from the deuterium in seawater.

FALLING WATER POWER

Prior to the Industrial Revolution, which began to take hold by about the middle of the eighteenth century, Europe and Great Britain were in danger of losing their forests which were their only source of fuel for stoking domestic and industrial fires. Also, scattered along rivers and streams were various types of waterwheels, used principally to provide power for mills.

There came a dramatic change with the invention of the steam engine in 1751, which became the prime mover of the Industrial Revolution, and the discovery of coal for which to fire it. With the marriage of steam and coal, large industrial centers sprang up near coal mines. Cities suffered from heavy air pollution created from the burning of large quantities of coal, and people were forced to live in the constant filth of coal ash and smoke. Lung ailments were common, particularly among the elderly. Despite this unpleasantness, however, the Industrial Revolution brought about prosperity and security never seen before.

Not all countries were as fortunate as Great Britain in having abundant coal resources for which to fuel their industries. By about the time the British were converting from wood to coal, the French were taking advantage of their so-called "white coal," or water power, which was their most abundant energy resource. The crude waterwheel gave way to more efficient hydraulic motors in order to achieve industrialization.

The new water motors were designed with horizontal waterwheels (FIG. 12-1) having curved blades, as opposed to the more traditional vertical waterwheels. The horizontal waterwheel rested at the bottom of a tall cylindrical chamber, and a tapered sluice supplied large quantities of water at an angle to the chamber so that water entered the chamber with an appreciable rotational velocity. The weight of the water above the wheel, along with the water's flow through the curved blades, made the wheel rotate through a combination of water pressure and rotational energy. This type of waterwheel was not particularly efficient, however, and the best efficiency it could achieve was only about 20 percent, which was considerably under that of some vertical waterwheels.

Waterwheels enjoyed a diversity of design, as well as application. The basic criteria was that the water must enter the hydraulic motor without impact and lose its initial velocity by the time it flows out of the motor. Therefore, no hydraulic energy should be wasted in the form of turbulence or unused kinetic energy. By the 1820s, efficient hydraulic motors were at work in the western world, some achieving as much as 70 percent efficiency. One reason their efficiency was so high was that their design was guided by scientific principles and not by the hit-or-miss approach generally used.

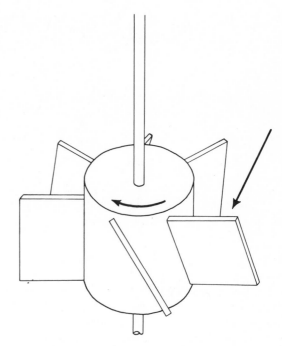

FIG. 12-1. A horizontal water wheel.

Out of these diverse designs came a fundamentally new kind of hydraulic motor, which henceforth made all other motors obsolete. This was the *water turbine*, a term coined by the French engineer Claude Burdin who provided the theoretical concept in 1811. His student, Benoît Fourneyron, designed and built over 100 industrial models that achieved 80 percent efficiency and more. One of his earliest turbines, built in 1832, had an 8-foot diameter and could generate 50 horsepower. Around 1843, a few of Fourneyron's turbines were sent to the United States, where they had a considerable influence.

Fourneyron's turbines were outward-flow models and could only maintain their efficiency under stringent operating conditions of water flow and load. If the water flow was reduced in order to reduce power output, the efficiency dropped sharply. Also, if the load on the turbine was suddenly removed and the turbine was not shut down immediately, it could race until it flew apart. This prompted the search for alternative designs, and by the 1840s, European engineers were putting their efforts into *axial-flow*, or propeller-type, turbines.

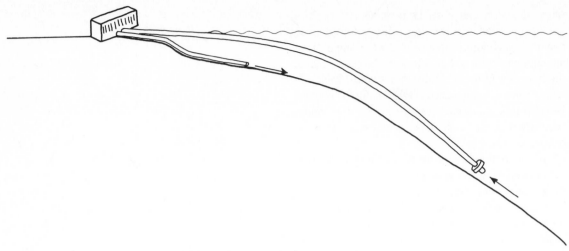

FIG. 12-2. An OTEC power station.

Meanwhile, in the United States, efforts were being applied toward inward-flow and mixed-flow turbines. With the inward-flow turbine, the water enters the motor from the outside through an array of fixed blades that direct the water onto the blades of the central rotor. The water then enters the race near the center of the rotor and flows away with most of its energy converted into the rotation of the turbine. The new turbines were small, ran at high speeds, worked submerged, used a wide range of heads of water, were more powerful, and had greater overall efficiency. When the electrical generator was invented and married to the water turbine in the late 1800s, together they made one of the most substantial contributions to the industrial age.

SEA POWER

The oceans are the world's largest solar collector. Every day, 30 million square miles of tropical seas absorb solar radiation with an equivalent heat content of 250 billion barrels of oil—more than the world's present total reserves of recoverable oil. If only a tiny fraction of this vast store of energy is converted into electricity, it will greatly enhance a country's future energy supply.

Alternative sources of energy are being pursued by several nations that do not want their national security jeopardized by heavy dependence on foreign sources of oil. This goal could be realized through a technology called ocean thermal-energy conversion (OTEC). If OTEC installations (FIG. 12-2) converted less than 0.1 percent of the energy stored as heat in the surface waters of the tropics, they could generate roughly 15 million megawatts (million watts) of electricity—more than 20 times the current generating capacity of the United States.

The OTEC system uses warm seawater to evaporate a working fluid that has a very low boiling point. In a *closed-cycle system*, a working fluid such as ammonia or Freon is enclosed in the system and recycles continuously. An example is a refrigerator. The warm seawater itself could also be boiled in a vacuum chamber, which lowers the pressure, thereby reducing the boiling point of water. This is an *open-cycle system*, in which the working fluid is a constantly changing supply of seawater. This system has the added benefit of producing desalinated water that could be used for irrigation in arid zones.

In both systems, the resulting vapor drives a turbine to generate electricity. Cold water drawn up from depths of 2000 to 3000 feet then condenses the working fluid to complete the cycle. The nutrient-rich cold water also could be used for culturing fish and serve nearby buildings with refrigeration and air conditioning. As long as there exists a temperature difference of about 40 degrees Fahrenheit between the warm surface water and the deep water, useful amounts of electrical power could be generated. The power plant could be located onshore, offshore, or on a mobile platform out at sea. The electricity could supply a utility-grid system, or it could be used locally for manufacturing substitute fuels such as methanol and hydrogen, for refining metals brought up from the seabed, or for making ammonia for fertilizer.

The idea of tapping thermal energy from the ocean was originally put forward by the French biophysicist Jacques d'Arsenval more than a century ago. He envisioned the use of a closed-cycle system but never tested his hypothesis. In 1930, a former student of his, Georges Claude, designed and built an open-cycle system at Matanzas Bay in northern Cuba. The system generated 22 kilowatts of electrical power, but unfortunately, it consumed more energy than it delivered, thereby producing a net loss of power. However, the experiment did demonstrate that cold water could be efficiently brought up to the surface from a depth of more than 2000 feet.

Claude's next major effort involved a floating open-cycle plant installed on a cargo vessel moored off the coast of Brazil. Unfortunately, the experiment was ruined after waves destroyed the large-diameter cold-water pipe as it was being deployed. Having invested his own money in these projects, Claude was on the verge of bankruptcy when he died before ever realizing his goal of generating net power from an open-cycle system.

The French government, which was highly interested in Claude's work, continued the research. In 1956, a French team designed a 3-megawatt plant for the Ivory Coast of Africa, but it was never built because of technical problems.

The energy crisis of the 1970s forced the United States and other nations to devote serious consideration to OTEC. The first prototype closed-cycle plant to produce a net output of electrical power was mounted on a barge moored off Keahole Point on the west coast of Hawaii. It generated 15 kilowatts of electricity. The U.S. Department of Energy tested a commercial-scale OTEC system installed on board a converted Navy tanker, which garnered energy as it slowly moved through tropical waters.

In 1981, the Japanese built a closed-cycle pilot plant that generated 35 kilowatts of power on the island of Nauru in the western Pacific.

The open-cycle system, also known as the *Claude cycle*, offers a number of advantages over the closed-cycle system. By using seawater as the working fluid, it eliminates the possibility of contaminating the marine environment with toxic chemicals. The heat exchangers of an open-cycle system are the direct-contact type, which are cheaper and more effective than those required for a closed-cycle system. Therefore, open-cycle plants should be more efficient at converting ocean heat into electricity, as well as less expensive to build. Direct-contact heat exchangers also could be made of plastic instead of metal; therefore, they would be less susceptible to corrosion and fouling in warm seawater. In addition to providing electricity, and open-cycle system can produce desalinated water as a byproduct.

The turbines of an open-cycle system would have to be up to 140 feet in diameter, considerably larger than the ones used in closed-cycle systems, because of the low density of the steam. Also, the vacuum chamber, which must provide low pressures equivalent to those above 15 miles altitude, would have to be quite large. Nonetheless, the technical problems with the open-cycle system probably will be solved by the time petroleum either becomes too costly or very scarce.

HARNESSING WAVES AND TIDES

The crash of a large wave as it breaks on the beach is a vivid example of the sizeable amount of energy that such a wave dissipates. The intertidal zones of rocky coasts receive much more energy per unit area from waves than they do from the Sun. Wave energy also helps enhance the diversity and productivity of marine organisms that live on wave-beaten shores between low and high tide.

Wave energy is an important contributor to the overall richness of coastal environments, especially coral reefs where the waves pound the hardest. The waves are generated by strong winds from distant storms blowing across large areas of the ocean. Local storms near the coasts provide the strongest waves, especially when they are superimposed on

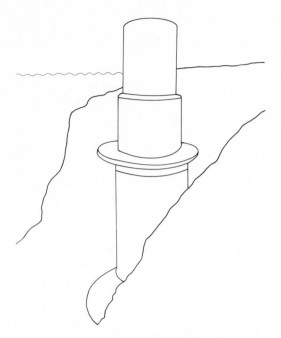

FIG. 12-3. The Norsk wave power generator
on the rocky coast of Norway.

the rising and falling tides. Efforts have been made
to put this abundant form of energy to use, such as
to generate electricity.

Norway, a country blessed with enough hydroelectric power to meet 100 percent of its present-
day needs, has turned to wave power because the
most efficient and economical hydroelectric schemes
have been developed. Therefore, as energy needs
expand, opening new hydroelectric projects will become ever more expensive. On the other hand,
wave power could be competitive within the next
decade.

The first such project was built by Kvaerner
Brug A/S and Norwave A/S at Toftestallen, near
Bergen, Norway, in the mid-1980s (FIG. 12-3). The
wave-powered generator represents the culmination
of more than 10 years of research and development.
Its basic operating principle is rather simple. A crashing wave at the base of the wave-powered generator compresses air and forces it into a vertical tower,
where the compressed air spins a specially designed
turbine called a *Pantor*. The whirling turbine in turn
drives an electrical generator, and current flows into

the grid of the Nord-Hordaland power board. The
wave-power station is expected to generate 1.8
gigawatts (billion watts) of electricity per year.

Gulfs, embayments, and certain sections along
the coast in all parts of the world have tides with
ranges exceeding 12 feet, called *macrotides*. The development of exceptionally high tidal ranges in certain embayments is a result of combinations of
convergence and resonance effects within the tidal
basin. As the tide moves into a narrowing channel,
the water movement is constricted and the resulting wedging augments the tide height. One method
of generating electricity using tidal power involves
damming an embayment, letting it fill with water at
high tide, then closing the sluice gates at the tidal
maximum when there is a sufficient head of water
to drive the water turbines.

Tidal power can save fuel that would otherwise
be burned in conventional power stations. It cannot
totally take the place of a conventional generating
station, however, because tidal power rarely coin-
cides with the ups and downs of electrical demand,
so spare generating capacity must be available when
the tides are out.

The La Rance Tidal Power Station (FIG. 12-4)
in northwestern France was completed in 1967. It
generates 65 megawatts of power throughout the
year. It has a 50-foot spillway, supplying over 1000
tons of water per second to four turbines. The project maximizes energy output by using excess energy
generated elsewhere in the system to pump water
at La Rance to even greater heights.

Studies in the Severn Estuary of the Bristol
Channel in southwestern England show that a
slightly more complex scheme than La Rance could
supply 12 percent of Great Britain's present electrical demand. Other macrotidal locations are generally too isolated for a tidal power station to be
economical and effective. Unless tidal power generation can be linked into an existing electrical grid,
the intermittent and inconvenient times at which
tides occur make it less attractive.

An alternate approach to tidal power generation
is to harness the kinetic energy of the water motion, as well as the potential energy associated with
the tidal height variations. Those locations that ex-

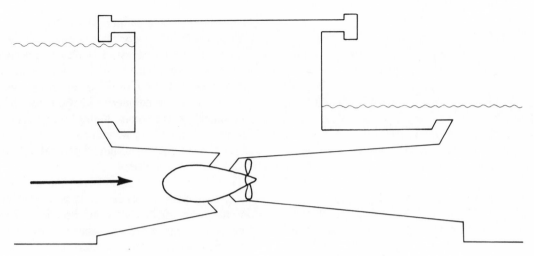

FIG. 12-4. Cross section of La Rance tidal power station.

perience macrotidal ranges also usually experience strong tidal currents. The idea is to use reversible-bladed turbines, which are rotated by both the incoming and outgoing seawater.

Macrotides are dependent on the shape of the bays and estuaries, which channel the wavelike progression of the tide and increase its amplitude. However, the construction of a confining structure such as a dam or barrage might alter the shape of the bay, which might in turn severely affect the enhancement of the tidal range. Some of these problems can be resolved by constructing a small-scale tidal model or by mathematical modeling on a computer.

The latest development in the ongoing search for efficient and economical ways to extract energy from moving water is an experimental hydroelectric generator that converts the to-and-fro surge of a tidal river, such as New York City's East River, into electricity. The three-bladed generator is about 15 feet in diameter and is attached to a river structure, such as a drawbridge (FIG. 12-5). It could produce around 20 kilowatts of electrical power during peak-velocity tidal flow, which for the East River is about 6 feet per second, or 4 miles per hour. The generated electricity is fed into the Con Edison power grid and monitored to determine the efficiency of the turbine.

This prototype hydroelectric generator is designed to operate only when the tide is flowing in one direction. Future designs will rotate to capture energy from the tide in both directions. The advantages of this type of generator is that is should be simple to install and maintain at a reasonable cost, and will compete favorably with power generated by coal and nuclear power plants.

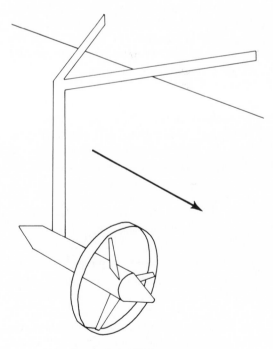

FIG. 12-5. Tidal generator.

A WINDY ALTERNATIVE

Wind power became an important source of energy when sailors harnessed it to propel their ships sometime around 3000 B.C. The first sailing ships used square sails, which meant they could only go in whatever direction the wind happened to be blowing and sailors trading goods at foreign ports had to wait until the wind reversed in order to return home. The use of a triangular sail attached to a pivoting boom let sailors tack against the wind, thereby allowing them to sail in any direction regardless of which way the wind was blowing. Unfortunately, sailors still were stranded in the doldrums, where the wind would not blow for long stretches.

The steam engine changed all that, and ships could come and go as they pleased in all kinds of weather. The steam engine gave way to the internal combustion engine, which gave way to the high-pressure steam turbine engine and the gas turbine engine for ship propulsion.

Now, because of the high cost of fossil fuel and dwindling resources, ship builders are returning to the sail—not ones made of canvas, but using entirely new concepts of gathering the wind. One is called the *Turbosail* and its inventor is the French oceanographer Jacques Cousteau. In June 1985, the experimental all-aluminum ship, *Alcyone*, successfully completed a trans-Atlantic crossing using two Turbosails as the main source of propulsion. These were hollow cylinders that could be oriented by computer control according to the wind direction. A fan at the top sucked air through a slot in the leeward side, generating lift, which translated into forward motion. The Turbosail wind-propulsion system was designed to cut fuel costs of commercial vessels by 15 to 35 percent. Other designs include fold-out sails that catch the wind when the ship is steaming with the wind and are tucked away when the ship heads into the wind.

One of the oldest methods of producing mechanical power was the windmill. The earliest windmills were in western Asia around 1000 A.D. and looked more like large paddle wheels mounted on a vertical shaft. They evolved into wood and canvas straight-bladed windmills mounted on horizontal shafts, which were used extensively for grinding grain, sawing timber, and pumping water, generally in areas where water power was not available.

Since the energy crisis of the 1970s, the windmill has been redesigned for more efficiency and greater power in order to drive electrical generators. As a result, a number of individual windmills and windmill farms are contributing to the local electrical power systems in several countries, thereby reducing the burden on fossil-fuel generating plants when the wind is blowing.

The major disadvantage of using the wind as an energy source is its unpredictability because of random changes in wind strength and direction. The location is also important, and seacoasts seemed to be the ideal places because of their steady onshore and offshore breezes, which are generated by the temperature difference between land and sea (FIG. 12-6). Unfortunately, such a location also places the windmill in danger from high winds generated by seasonal storms. The new designs must be able to cope with these problems by taking advantage of some unique properties of the wind.

The energy of the wind passing through a certain area at ordinary wind velocities is comparable with the Sun's radiation on an equal area. This makes wind a highly diffused form of energy, and its drag force is quite small compared to that of other forces, such as flowing water. When the wind is blowing at a brisk pace of 10 to 15 miles per hour, the energy flux is something on the order of about 150 watts or about one-quarter horsepower per square yard of surface area. Therefore, in order to generate a significant amount of power, a windmill must harvest a large cross-sectional area of the wind.

Aerodynamic lift forces can produce several times the output of drag forces, wherein the blades of a windmill are simply dragged along by the same cross-sectional area of the wind. Moreover, a windmill taking advantage of lift needs less blade area than one based on drag because lifting blades do not need to cover as much area with their surface, but merely sweep across the wind repeatedly. Therefore, the blades rotate faster and more efficiently.

The windmill's size is also an important factor. The larger the windmill, the more cost-effective it becomes because, among other things, it can do

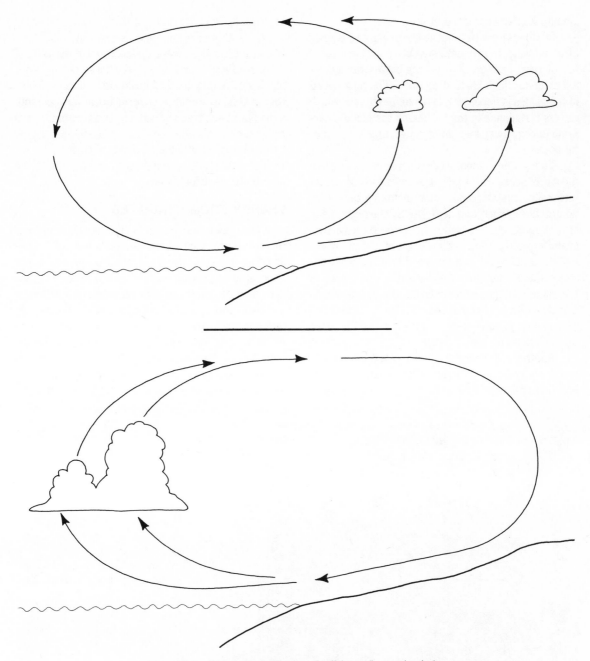

FIG. **12-6.** The onshore (top) and offshore (bottom) winds.

more work with the same number of personnel. Today's experimental wind turbines have reached about 300 feet in diameter, and even larger ones would be desirable for generating electricity in sig-

nificant amounts. However, increasing the size also increases the danger from atmospheric turbulence and structural costs, which eventually outweigh the economic gains. Also, metal fatigue on long blades inhibits further increases in size.

In order to optimize its ability to capture energy from the wind, the windmill should have a large number of long, slender, fast-moving blades whose airfoil exhibits a high lift-to-drag ratio. The high speed of the blades is necessary to minimize torque, which causes turbulence in the air behind the blades, carrying energy away instead of converting it into useful power.

Today's new windmill designs look nothing like their predecessors. A tall, eggbeater-looking structure called the Darrieus rotor, named after its inventor the French engineer J.G.S. Darrieus, uses two or more curved blades mounted on a vertical shaft. The advantage of this design is that the windmill does not need to be pointed into the wind because the air currents always strike the blades at the same attitude, causing lift on the airfoil, which translates into rotational energy. Also, all mechanical and control equipment is at ground level for ease of operation and maintenance.

Another giant windmill looks as though the engineers left off half of the propeller. The single blade is counterbalanced by a weight in place of the other blade. This weight reduces blade area, increases speed, and lowers the cost of the windmill.

A futuristic concept envisions planting a large number of giant vertical-axis windmills, looking much like huge rotating football goalposts, on the sea bottom in shallow offshore waters. Locating the windmills out at sea takes advantage of the constant sea breeze, as well as reduces their noise impact. Great Britain is looking into the feasibility of such a scheme, which could make an important contribution to the country's electrical needs.

ENERGY FROM SEAWATER

Thermonuclear fusion is the energy source of the Sun. In the Sun's core, temperatures of 10 million degrees Celsius and extremely large pressures generated by gravitational forces are needed to produce fusion—conditions that only exist on Earth during the detonation of a hydrogen bomb. Therefore, one approach toward producing a controlled thermonuclear fusion reaction is to zap tiny hydrogen bombs in the form of pellets of deuterium and tritium (another heavy isotope of hydrogen) with high-temperature lasers (FIG. 12-7) or beams of high-

(Courtesy of U.S. Department of Energy)

FIG. 12-7. Installation of mirrors for laser fusion research at Los Alamos, New Mexico.

energy lithium ions. This would compress the hydrogen fuel 1000 times and heat it 100 million degrees Celsius, causing a small, controlled, thermonuclear explosion.

The amount of energy released from thermonuclear fusion is tremendous. According to Albert Einstein's special theory of relativity, the extra mass left over when the two hydrogen isotopes are fused into helium is converted into a prodigious amount of energy. Just one ounce of hydrogen consumed in this manner yields the energy equivalent of about 80,000 gallons of gasoline, or 10 million times more energy than if the hydrogen were simply burned.

For more than 30 years, physicists around the world have relentlessly pursued the goal of thermonuclear fusion. In a world of rapidly diminishing fossil fuels, an energy source that is both renewable and essentially nonpolluting is worth this effort. The fuel for fusion is abundantly available in seawater, and the energy from the fusion of deuterium, a heavy isotope of hydrogen, in a pool of water 100 feet on each side and 7 feet deep could provide the electrical needs of a city of 0.25 million people for a year. Fusion is safe, and its byproducts are energy and helium, a harmless gas that escapes into space.

However, the long haul to develop controlled thermonuclear fusion has been plagued with failure and frustration. Many milestones have been passed, but real success always seems to be at least 20 years into the future. Because of the enormous engineering and technical problems that still remain, some experts question whether energy from thermonuclear fusion will ever be economically viable.

13

Harvesting the Sea

BIOLOGISTS of the Smithsonian Institution using a deep-sea submersible made a remarkable discovery off the Bahamas that might open up a whole new realm of oceanography. On an uncharted seamount, the scientists found a totally new and unexpected plant at a depth of about 900 feet—deeper than any previously known oceanic plant larger than a microorganism. It previously had been thought that the deepest depth at which plants could survive was 600 feet because very little sunlight penetrates below that depth.

The new plant is a variety of purple algae with a unique structure, consisting of heavily calcified lateral walls but very thin upper and lower walls, which enable it to make the most of the feeble sunlight. The cells are stacked on top of one another like a stack of cans at a grocery store. The discovery will force scientists to rethink the role that such algae play in the productivity of the oceans, marine food chains, sedimentary processes, and reef building.

VARIETY IS THE SPICE OF LIFE

The oceans have far-reaching effects on the climate and the composition and distribution of life. They also have profound effects on the course of evolution, and therefore the history of life. The chief mechanism by which evolution proceeds is *natural selection*, which is mainly an ecological process based on the relationship between organisms and their environment. Certain inherited traits allow species to become particularly well suited to survive and reproduce in their prevailing environment. During environmental change, species with favorable traits adapt more easily and are more likely to survive and pass on their "good" genes to their offspring. Because there are a number of different environments, this process results in a wide variety of species, and species that have the same genetic makeup but that are oceans apart might have entirely different physical and behavioral characteristics.

Variability in the oceans is greatly affected by the ocean currents, the temperature, the nature of seasonal fluctuations, the distribution of nutrients, the patterns of productivity, and many other factors of fundamental importance to living organisms. Therefore, evolutionary trends in marine animals varied through geological time in response to major environmental changes as natural selection acted to adapt organisms to the new conditions.

The vast majority of marine species live on the continental shelves or on shallow-water portions of islands and subsurface rises at depths less than 600 feet. The richest shallow-water faunas are found at low latitudes in the tropics, which are crowded with vast numbers of highly specialized species. Progressing to higher latitudes, diversity gradually falls off, until in the polar regions, there are less than one-tenth as many animals as there are in the tropics.

The diversity depends on the stability of the food supply, and as seasons become more pronounced, as they do in the higher latitudes, there are greater fluctuations in food production. Longitude also affects diversity, which in turn is affected by seasonal changes such as variations in surface and upwelling currents that affect the nutrient supply, causing large fluctuations in productivity. Zones of cold, nutrient-rich upwelling water cover only about 1 percent of the ocean but account for about 40 percent of the production. Modern fishermen use sophisticated tools such as satellites to track down these areas of upwelling water, which are generally where the fish are.

The greatest diversity among species is off the shores of small islands (FIG. 13-1) or small continents in large oceans, where fluctuations in nutrient supplies are least affected by the seasonal effects of landmasses. The least amount of diversity is off large continents, particularly when they face small oceans where shallow-water seasonal variations are the greatest. Diversity also increases with distance from large continental landmasses.

Diversity is highly dependent on the stability of food resources, which in turn depend largely on the shape of the continents, the extent of inland seas, and the presence of coastal mountains. Continental platforms are of particular importance because not only do extensive shallow seas provide much habitat area for shallow-water faunas, but also such seas tend to dampen seasonal climatic changes and make the local environment more amenable.

Species in one ocean tend to be quite different from their cousins living in different oceans or on opposite sides of the same ocean. Even along a continuous coastline, there are major changes in species from place to place that generally correspond to climatic changes. This is because latitudinal and climatic changes create barriers to shallow-water organisms. The deep-sea floor provides another formidable barrier to the dispersal of shallow-water organisms. Also, the midocean ridges form a series of north-south barriers to shallow-water marine organisms.

As the result of these barriers, the marine faunas are partitioned into more than 30 individual provinces. In each province, there is generally only a small percentage of common species. The shallow-water marine fauna represents more than ten times as many species as would otherwise be present in a world with only a single province. Such conditions might have existed 200 million years ago when there was only one supercontinent and a single large ocean.

By far, the widest ranging marine province as well as the most diverse, is the Indo-Pacific province as a result of its long chains of volcanic island arcs. When long island chains are arranged east to west within the same climatic zone, they are inhabited by wide-ranging faunas that are highly diverse. The fauna spill from these arcs onto adjacent tropical continental shelves and islands. However, this vast tropical biota is cut off from the western shores of the Americas by the East Pacific Rise, which is an effective obstruction to the migration of shallow-water organisms.

THE COLD AND THE DARK

In the Arctic Ocean, sea ice inhibits primary production for 8 months of the year (FIG. 13-2). During the winter, algae trapped in the sea ice at freeze-up show only slow increases in abundance, followed by rapid proliferation when light returns in the early spring. In such an environment, many species of tiny, plant-eating crustaceans called *copepods*, which are an important link in the food chain, require 2 or more years to complete their life cycles. Reproduction usually takes place just before or immediately after the breakup of the ice. Development is relatively rapid during the open-water season from April to September, with most individuals reaching adolescence before the onset of new ice. The copepods overwinter beneath the ice from October until mid-May, emerging as mature adults in late winter and early spring. During the winter, the hungry copepods migrate toward the ice, where they graze on ice algae melted at the interface between the bottom of the ice and the seawater.

The seas surrounding Antarctica, which together are the coldest marine habitat in the world, were once thought to be totally barren of life. In 1899, a British expedition to the southernmost continent found examples of previously unknown fish species related to the perchlike fishes common to many parts of the world. Upwards of 100 species of the fish are confined to the Antarctic region, accounting for about two-thirds of the fish species in the area. Living in subfreezing waters, the fish rely on an antifreeze like substance in their bodies to keep from freezing. The fish also rely on neutral buoyancy, or weightlessness, which spares them from having to expend precious energy to stay afloat,

(Courtesy of NOAA)

FIG. 13-1. A coral garden near the Virgin Islands.

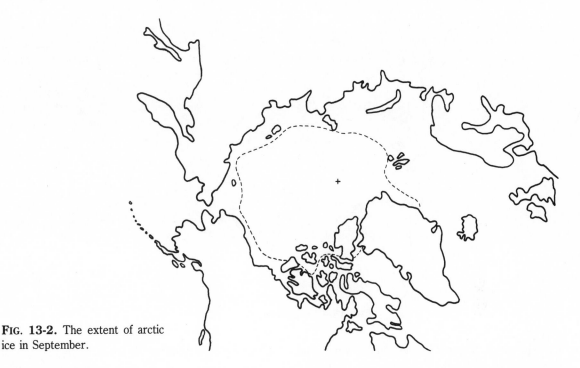

FIG. 13-2. The extent of arctic ice in September.

especially in the Antarctic winter when the food supply is scarce.

The Antarctic seas are cut off from the general circulation of the ocean by a circum-Antarctic current, which becomes a thermal barrier, impeding the inflow of warm currents and warm-water fishes, as well as the outflow of Antarctic fishes. Also, because of the extreme cold and certain other factors, the Antarctic seas have less diversity than the Arctic Ocean, which supports almost twice as many species.

For 4 months of the year, Antarctica is in total darkness. Even in the short summer, the water under the ice receives less than 1 percent of the sunlight that strikes the surface. More hazardous for the fishes than the cold and the dark is the danger of ice. The sea ice covers the water for at least 10 months of the year (FIG. 13-3). The water temperature throughout the year varies only from about 2 to 4 degrees Fahrenheit below freezing. Since fish are cold blooded, their body temperature is essentially the same as their environment. As long as no ice enters their bodies to cause ice crystals to propagate and freeze body fluids, most fish can en-

dure when their blood cools as much as 2 degrees Fahrenheit below freezing. Their endurance is mainly a result of the salt content, along with other antifreeze-like substances in their blood. Antarctic fishes will freeze when their temperature reaches approximately 28 degrees Fahrenheit.

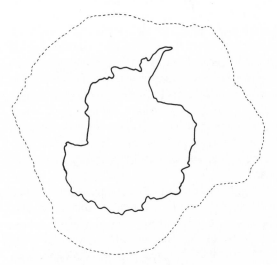

FIG. 13-3. The extent of drift ice in Antarctica.

FIG. 13-4. The major coral reefs.

The deep waters of the ocean were once thought to be a vast, barren desert. In the 1870s, the research vessel H.M.S. *Challenger*, while dredging on the ocean bottom, brought to the surface a vast collection of bottom-dwelling creatures even from the deepest trenches. The inventory included some of the most bizarre forms ever molded by adaptation to extreme conditions, along with several animals that were thought to have been long extinct. More recently, a population of large, active animals were found thriving in total darkness as deep as 4 miles, in what was thought to be the domain of small, feeble creatures such as sponges, worms, and snails that were specially adapted to live off the debris of dead animals raining down from above. Much of the deep-sea floor was found to be teaming with numerous species of scavengers, including highly aggressive worms, large crustaceans, deep-diving octopuses, a variety of fishes, and giant sharks.

Large numbers of fish existing at great depths in the lower latitudes are related to commercially harvested varieties living in the higher latitudes. Some species that are generally considered arctic fishes apparently represent near-surface expressions of populations that inhabit the cold, deep waters off

continental margins to the south. If this is true of other coasts as well, the size and extent of a number of fish populations might be considerably larger than it is thought. The total numbers of large shrimps in particular must be immense since they were found in all regions at all depths.

The uniformly large size of most deep-water fish could indicate that they are all adults. It is probable that the juveniles of these fishes inhabit much shallower depths and descend when they mature. The abundant bottom-dwelling animals constitute an entirely new fishery resource. Underdeveloped countries will particularly want to harvest these potential deep fisheries off their coasts.

THE CROWDED SEAS

Among the builders of the world, coral rivals even man's ability to alter the surface of the Earth. Over eons of geologic time, coral polyps have accreted massive reefs of limestone. The reefs are limited to clear, warm, sunlit tropical oceans such as the Indo-Pacific and the Western Atlantic (FIG. 13-4).

Coral reefs are important land builders in the tropics, forming entire chains of islands and altering the shoreline of continents. Hundreds of *atolls*, which are rings of coral islands, enclosing a central lagoon dot the Pacific. They consist of reefs several thousand yards across, many of which are formed on ancient volcanic cones that have subsided below the sea, with the rate of coral growth matching the rate of subsidence.

Fringing reefs grow in shallow water and closely border a coast or are separated from it by a narrow stretch of water. *Barrier reefs* also parallel a coast but are farther out, are much larger, and extend for greater distances. The best example is the Great Barrier Reef off the northeastern coast of Australia. It forms an underwater embankment more than 1200 miles long, 90 miles wide and 400 feet high.

The coral-reef environment supports a larger number of plant and animal species than any other habitat (FIG. 13-5). The key to this prodigious productivity is the unique biology of corals, which plays a vital role in the structure, ecology, and nutrient cycles of the reef community. Coral-reef environments have among the highest rates of photosynthesis, nitrogen fixation, and limestone deposition of any environment. The most remarkable feature of coral colonies is their ability to form a massive calcareous skeleton, weighing several hundred tons and large enough to fill a living room.

The coral polyp itself is soft-bodied and is essentially a contractible sac, crowned with a ring of six tentacles tipped with poisonous stingers that surround a mouthlike opening. The polyps live in individual skeletal cups composed of calcium carbonate. They extend their tentacles to feed by night and withdraw into the cups by day or during low tide in order to keep from drying out in the sun. The corals live in symbiosis with a certain algae

(Photo by P.E. Cloud, courtesy of USGS)

FIG. 13-5. A collection of sea life on the ocean floor.

known as zooxanthellae, which live within their host's bodies where they take in waste products and give off organic matter that is absorbed by the coral. Because the algae need sunlight for photosynthesis, corals live only in warm ocean waters less than 300 feet deep, with much of the coral growth being within the intertidal zone.

Coral polyps might not always dominate the living organisms, nor all the biological activity in all parts of the coral reef. Nonetheless, the existence of many plant and animal communities of the coral reef is based on the coral's ability to build massive, wave-resistant structures.

The major structural feature of a living reef is a coral rampart that reaches almost to the surface of the water. It consists of massive rounded coral heads and a variety of branching corals (FIG. 13-6). Living on this framework are smaller, more fragile corals and large quantities of green and red calcare-

ous algae. Hundreds of species of encrusting organisms, such as barnacles live on top of the coral framework. Large numbers of invertebrates and fishes hide in the nooks and crannies of the reef, some of them emerging only at night. Other organisms are attached to virtually all the available space on the underside of coral plates or on dead coral skeletons. Filter feeders such as sponges and sea fans occupy the deeper regions.

The fore reef is seaward of the reef crest and blankets nearly the entire seafloor. In deeper waters, many corals grow in flat, thin sheets to maximize their light-gathering area. In other areas, the corals form massive buttresses separated by narrow sandy channels composed of calcareous debris from dead corals, calcareous algae, and other organisms. The channels resemble narrow, winding canyons with vertical walls of solid coral. They dissipate wave energy, allowing the free flow of sedi-

(Photo by K.O. Emery, courtesy of USGS)

FIG. 13-6. Coral at Bikini Atoll.

ments that would otherwise choke the coral.

Down from the fore reef is a coral terrace, followed by a sandy slope with isolated coral pinnacles, then another terrace, and finally an almost vertical drop into the darkness of the abyss. The rise and fall of the sea level during the last few million years has produced terraces that resemble a stair-step growth of coral, running up an island or a continent. The drowned coral represent periods of extensive glaciation, which caused the sea level to drop dramatically. In Jamaica, almost 30 feet of reef has built up since the present sea level stabilized some 5000 years ago, following the last ice age.

Because coral reefs are centers of high biological productivity and their fishes are a major source of food in tropical areas, the spread of tourist resorts along coral coasts in many parts of the world might adversely affect the production of these areas. Such developments are almost always accompanied by increased dumping of sewage; by overfishing; by physically damaging the reef through building, dredging, dumping, and landfilling; and by destroying the reef on a large scale to provide tourists with souvenirs and curios. In many areas, such as Bermuda, the Virgin Islands, and Hawaii, development and sewage outflows have led to extensive overgrowth and killing of the reef by thick mats of algae. These algae suffocate the coral by supporting the growth of oxygen-consuming bacteria, particularly in the winter when the algal cover on shallow reefs is very high. The result is that the living coral die and the reef is eventually destroyed.

THE KRILL KILL

After the depletion of the whales in the northern seas, whalers turned their attention to the southern oceans. Modern whaling in the Antarctic began around 1930 with the advent of the factory ship, which was a self-contained floating plant that could stay at sea for months on end, steadily catching and processing whales. The whales were seldom caught for their meat, but mainly for their blubber which rendered a fine oil used in making perfume and other products. By 1935, there were fears that whales in the Antarctic would be depleted as they had been in the northern seas, while at the same time, a sur-

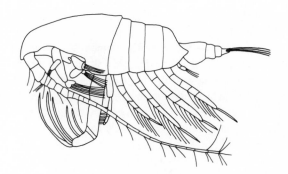

FIG. 13-7. Krill.

plus of whale products flooded the market, which drove prices down. This prompted the formation of the International Convention for the Regulation of Whaling.

The whales were given a break during World War II. After the war when the whaling resumed, the International Whaling Commission was formed with the aim of managing whaling in the southern oceans and elsewhere. Its disregard of warnings from the scientific community that whales were being exploited too heavily and pressures from various governments and from international environmental organizations forced the Commission to adopt new management procedures in 1974. Under the new rules, the world's oceans were partitioned into several sectors where whales were classified either for protection or for exploitation, and catch quotas were set for commercial whaling. Despite the Commission's international ban on all commercial whaling imposed in 1986, several countries, particularly Japan, Norway, and the Soviet Union, have continued extensive whaling under the guise of scientific research. A ban on scientific whaling is virtually impossible to enforce.

In the southern oceans, whales, fishes, squid, and sea birds feed on a small, shrimplike crustacean called *krill* (FIG. 13-7). With the decline of whale populations, other krill-eating animals in the southern oceans have shown population increases in recent years. For instance, antarctic seals have had population increases that have outpaced any simple recovery from past overhunting. Similarly, the population of penguins seems larger than can be explained simply as recovery from the slaughter of the nineteenth century. Population increases also have been documented for some sea birds.

Apparently, the removal of most of the large baleen whales (whales that strain krill through baleen plates) from the southern oceans has caused vast competition by other species for the krill that would have otherwise been eaten by the whales. Therefore, the depletion of the whales has resulted in a surplus of krill. The increase in krill has in turn attributed to the increase in populations of other krill-eating species.

By harvesting organisms further down the food chain (FIG. 13-8), mainly for their protein content, man has been placed in an unusual position of directly competing with other predators for this particular food supply, rather than just harvesting the predators themselves. The krill run in dense schools, and a cubic yard of seawater contains about 50 pounds of krill. Roughly 1 million tons of krill are harvested each year, using factory ships that freeze the catch at sea.

Unfortunately, very little of the krill is eaten directly by human consumers; most is ground up into animal feed for beef cattle and other domestic animals. At a 7 to 1 conversion rate—it takes 7 pounds of krill to make 1 pound of beef—this seems to be a tragic waste of natural resources. In order to maximize the sustainable yield for krill, whales and other animals would need to be eliminated.

Although the maximum landed tonnage of protein might be realized by eliminating whales and exploiting only the remaining krill, this solution would be unethical, as well as unaesthetic. Ultimately, the rules must take into account the social and political considerations, along with the biological ones. Because the relationships among species of the southern oceans is poorly understood, some guidelines are needed in order to maintain an ecological balance of antarctic marine living resources.

There are various other examples among the world's fisheries where major fish supplies have collapsed. In the tropics, a substantial fishing industry in the Gulf of Thailand has seen changes in the relative abundance of various species. As a result, there is a decline of roughly tenfold in the catch of "good" fish as compared with "trash" fish. The composi-

FIG. 13-8. The marine food chain.

tion of the catch is changing toward smaller fish species, and even the average size of fish within the same species is smaller. Thus, some good fish are now regarded as trash fish because only small individuals are caught. Overfishing tends to drive the populations below the levels at which competition is important in regulating population densities. Therefore, under heavy exploitation, species that can produce offspring quickly and copiously are given a relative advantage, and the fleshy, good fish (FIG. 13-9) give way to the coarser trash fish.

Beginning in 1960, the total fish catch from the North Sea doubled from 1.5 million tons to 3 million tons in 7 years. Since 1967, the catch has remained steady, with a pronounced decline in heavily exploited herring and mackerel, which is compensated by increased yields of cod, haddock, pollack, and whiting, along with other small fishes that are considered trash fish. It is not certain to what extent these changes are a result of shifts in fish popula-

tions, changes in patterns of commercial fishing or environmental effects.

It has been suggested that a systematic removal of large predator fish could allow annual catches of 5 million tons. However, such catches would consist of smaller trash fish, which would eventually dominate the North Sea. There, as well as off the coast of New England, population changes tend to be more variable and unpredictable than in the Gulf of Thailand because of the strongly seasonal behavioral pattern of the fishes and the significant differences in climate and other environmental conditions from one season to the next.

FARMING THE WORLD'S WATERS

One of the best examples of what modern aquaculture can achieve is the shrimp industry. In the past decade, world shrimp production has nearly doubled to 2 million tons. Pregnant female shrimp are caught off Malasia and flown to Japan, where they

(Courtesy of NOAA)

FIG. 13-9. Skipjack tuna.

(Courtesy of USDA)

FIG. 13-10. Harvesting catfish near Tunica, Mississippi.

TABLE 13-1. Productivity of the Oceans.

LOCATION	PRIMARY PRODUCTION TONS PER YEAR OF ORGANIC CARBON	PERCENT	TOTAL AVAILABLE FISH TONS PER YEAR OF FRESH FISH	PERCENT
Oceanic	16.3 billion	81.5	.16 million	0.07
Coastal Seas	3.6 billion	18.0	120.00 million	49.97
Upwelling Areas	0.1 billion	0.5	120.00 million	49.97
Total	20.0 billion		240.16 million	

fetch as much as $400 apiece—not to be eaten but to be used as breeding stock. A single female tiger shrimp can lay as many as 100,000 eggs, becoming mother to several tons of shrimp when they reach maturity. Such a price for a creature that only weighs a few ounces demonstrates the rising significance of what is being called *mariculture*, or the commercial raising of marine species.

Although the domestication of marine life has a long tradition in Asia, it is rare in Europe. There has been a modest beginning in West Germany, with the Baltic Sea Trout Aquaculture Society. Because fish grow fastest in warm water, biologists use the waste heat from a power plant to warm fish-farm waters. The company plans to harvest as much as 35 tons of valuable fish. Whether the venture will be profitable remains to be seen. The German Agricultural Ministry warned in 1984 that it would be a mistake to think that aquaculture can make up for declining fish catches in the ocean, at least in the foreseeable future.

Valuable sea animals are being raised increasingly for human consumption (TABLE 13-1). The shrimp, lobster, eel, and salmon now being raised by aquaculture account for less than 2 percent of the world's annual seafood harvest of roughly 75 million tons. Their total value is estimated to be five to ten times greater, however.

Norway is taking advantage of the warm Gulf Stream current, protected coastal areas, and fjords for intensive fish production. The country earned nearly $120 million in 1983 by exporting home-raised salmon. A Norwegian firm has not stopped there and offers a ready-made fish farm designed to produce 7000 tons of fish per year.

Experts consider the development of aquacul-

ture and mariculture to be another way of meeting the world's growing need for food. The Chinese lead the world and have more than 25 million acres of water in the form of canals, ponds, reservoirs, and natural and artificial lakes all stocked with fish. This is one method they have used to abolish hunger in their country.

World food needs also might be met by cultivating seaweed and algae, which are coming into their own as sources of nourishment. Twenty edible kinds of seaweed are gathered in Japan, making the Japanese, who eat about 1 pound of dried algae preparations weekly as appetizers or desserts, the world's leaders in consuming sea plants. The seaweed is rich in vitamins, and some of it is harvested wild, but many varieties are also cultivated.

When algae is grown under controlled conditions, it multiplies rapidly and produces large quantities of plant material that can be used as food. Algae crops can be harvested every 3 days, whereas a crop like corn requires 9 to 13 weeks between planting and harvesting. An acre of water could yield 30 tons of algae per year, as opposed to an average of 1 ton of wheat per acre of land.

The algae can be artificially flavored to taste like meat or vegetables, and because it contains over 50 percent protein, it would be just as nutritious. A new type of white algae has a naturally pleasant flavor, and when dried, it resembles flour and can be baked into bread or cake. One company makes more than 70 algae products for more than 100 uses in printing textiles, food processing, and cosmetics. This goes to prove that the ocean farm is immensely rich and can meet much of the human nutritional needs far into the future, provided man does not turn it into a desert as he has done with so much of the land.

14

A Sea Change

MODERN technology has given man the tools of measurement and computation so he can study the Earth as a system (FIG. 14-1). Scientists now can gain comprehensive knowledge of the state of the earth and of its global processes. They also have become uncomfortably aware that changes are taking place and that our own species is responsible for much of these changes. Economic developments over large portions of the Earth have required dramatic changes in the patterns of land and water use. There has been large-scale extraction of energy from fossil fuels and widespread use of man-made chemicals in industry and agriculture. These activities are believed to be related to alterations in cycles of essential nutrients. They appear to also affect the climate, with altered precipitation patterns worldwide.

Humans currently consume directly and indirectly some 40 percent of the terrestrial *net primary production* (NPP), which is the energy trapped by photosynthetic organisms throughout the globe. By doubling the human population during the early part of the next century, our species would require 80 percent of the world's NPP. To some ecologists, this is a preposterous notion, considering the deadly impacts of today's level of human activities. They are implying that human populations are growing so explosively and are modifying the environment so extensively that they are inflicting a global impact of unprecedented dimensions.

DEATH IN THE SEA

Throughout geologic history, there have been vast numbers of vanishing species. During geologically brief intervals of several million years, mass extinctions in the ocean have eliminated most of the species and half the families of plants and animals. Devastation of this magnitude could have been inflicted only by radical changes in environment on a global scale (FIG. 14-2). A number of theories have been put forward to explain the phenomenon, including cosmic radiation from a nearby exploding supernova or a massive meteor bombardment. There might have been drastic changes in the environmental limiting factors, including temperature and living

166

space on the ocean floor which determine the distribution and abundance of species in the sea.

The most important factor limiting the geographical distribution of marine animal species is water temperature. A particular species such as coral can only survive within a narrow temperature range (FIG. 14-3). An episode of climatic cooling could extinguish any species that could not adapt to the new, colder temperature, or find warmer refuge in which to migrate.

The first major mass extinction took place in the late Precambrian era, about 650 million years ago. At that time, animal life was still sparse and extinction decimated the ocean's population of single-celled phytoplankton, which were the first organisms to evolve cells with nuclei. The mass disappearance of this species coincided with a period when glaciers covered many parts of the Earth. When the ice disappeared near the end of the Precambrian, there was an explosion of species.

There was another mass extinction at the end of the Ordovician period, about 440 million years ago, which eliminated some 100 families, or about 6000 species of marine animals. Another major extinction occurred near the end of the Devonian period, about 370 million years ago, in which many tropical marine groups simultaneously disappeared. The largest mass extinction occurred at the end of the Paleozoic era, about 240 million years ago, when upwards of 95 percent of all the marine species vanished. The most well-known extinction was that of the dinosaurs along with 70 percent of all species at the end of the Cretaceous period, about 65 million years ago.

The oldest species alive in the world's oceans today are the ones that thrive in cold waters. Many arctic species, including certain brachiopods, star fishes, and bivalves, belong to biological orders whose history extends hundreds of millions of years to the Paleozoic era. In contrast, tropical faunas, such as reef communities, which were battered by periodic mass extinction have come and gone quite rapidly on the geologic time scale.

Even animals that shared the same environments were not identically affected, however. Whatever were the agents of biological stress—either climatic changes, shifts in ocean currents, shallowing seas, or disruptions in food chains—the ability

FIG. 14-1. A buoy designed to collect and transmit ocean and atmospheric data to shore stations.

(Courtesy of NOAA)

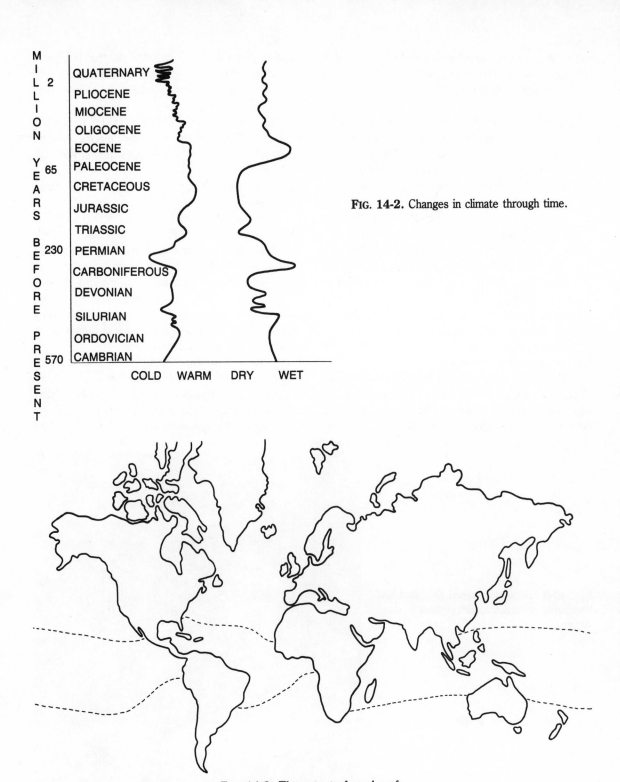

FIG. 14-2. Changes in climate through time.

FIG. 14-3. The extent of coral reefs.

of the biosphere to resist them was varied in different parts of the world. One very consistent pattern of mass extinctions, however, is that although each event typically affects different suites of organisms, tropical biotas, which contain the highest number of species, were nearly always the hardest hit.

ALTERED STATES

Not even the largest and most prominent features of the Earth's surface, such as mountains and seas, can be regarded as permanent. Because the rigid lithologic plates that make up the surface are in motion, continents and oceans are constantly being reshaped and rearranged. Mountain building associated with the movement of plates alters patterns of river drainage and climate. Examples of these changes are found extensively in the geologic record. One is when the Mediterranean Sea completely dried up 6 million years ago. For almost 1 million years, the seafloor remained a desert basin 1¼ miles below the surrounding continental plateaus. When either the Strait of Gibralter subsided or the sea level of the Atlantic rose, the basin was flooded again. It took several centuries to refill the Mediterranean. During that time, there existed what must have been the most spectacular of all waterfalls.

The Black Sea had a similar fate. Like the Mediterranean, it is a remnant of an ancient equatorial ocean called the *Tethys Sea*, which separated Africa from Europe and connected the Atlantic with the Indian Ocean. A collision of the African plate with Europe and Asia 20 million years ago squeezed out the Tethys, resulting in a long chain of mountains and two inland seas. One was the ancestral Mediterranean and the other was a composite of the Black, Caspian, and Aral seas, called the *Paratethys Sea*, which covered much of Eastern Europe. About 15 million years ago, the Mediterranean and the Paratethys separated, and the Paratethys became a brackish sea, much like the Black Sea of today.

The disintegration of the great inland waterway was closely associated with the sudden drying up of the Mediterranean. In a brief moment (geologically speaking), the Black Sea became almost dry. Then during the last ice age, it refilled and became a freshwater lake. The brackish and largely stagnant sea

that occupies the basin today has evolved since the end of the last ice age, and is not much older than civilization.

The Dead Sea is the lowest place on the face of the Earth: about 1300 feet below sea level. For thousands of years, freshwater carrying salts leached from rock, sand, and soil has flowed south through the Jordan Rift Valley into the terminal lake. Because there is no outlet, the inflowing water evaporates into the desert air, leaving salts to accumulate in the lake. As a result, the Dead Sea is the world's deepest and saltiest natural lake, with an average salinity eight times greater than the ocean. The hypersaline water is not particularly hospitable to life, hence the sea's name.

Since the 1960s, the diversion of freshwater from rivers feeding the Dead Sea for agricultural purposes lowered the lake and made its surface water saltier and much denser. The instability caused the lake to completely turn over, with the salty surface water trading places with the deeper water below. The changes in salinity had a major impact on the Dead Sea's sparse biota, and after the overturn, the lake seemed to be totally sterilized. Then in the rainy winter of 1979-80, the number of organisms increased dramatically. The Dead Sea was not entirely dead yet.

THE CARBON CONNECTION

Since the Industrial Revolution, human activities have increased the amount of carbon dioxide in the atmosphere from 0.027 percent to 0.035 percent—an increase of about one-third. If present trends continue (FIG. 14-4), by the year 2020 the amount of atmospheric carbon dioxide could be twice the current value, and global mean surface temperature could increase by about 5 degrees Fahrenheit. An increase in the average world temperature could change worldwide precipitation patterns and enlarge the area of the arid zones, significantly affecting agriculture.

The increase in carbon dioxide is attributed to the burning of fossil fuel and to the global destruction of forests. Although carbon dioxide only exists in trace amounts in the Earth's atmosphere, it plays a critical role in controlling the climate of the Earth

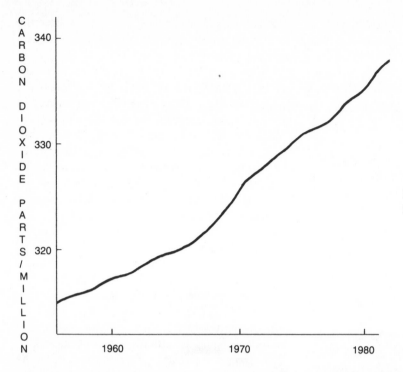

FIG. 14-4. The increase in atmospheric carbon dioxide.

by absorbing radiant energy. Excess heat trapped in the atmosphere has a large potential for substantially altering the world's climate. Carbon dioxide also plays a critical role as a source of carbon, which is fixed by photosynthesis in green plants and therefore provides the basis for all plant and animal life.

The burning of fossil fuels on the scale it is being done today is equivalent to releasing into the atmosphere 1 ton of carbon (in the form of carbon dioxide) for each of the world's 5 billion people every year. Americans alone release about 6 tons of carbon per person per year. At present, the atmosphere holds about 700 billion tons of carbon, and the increase of carbon released by burning fossil fuel alone is approximately 0.7 percent annually. Some of this carbon is taken out of the atmosphere by biological, hydrological, and geological processes, so that the average annual increase of atmospheric carbon dioxide is a little less than 0.5 percent.

The consumption of fossil fuels is expected to continue to increase well into the next century. When stocks of petroleum begin to run out, they will for the most part be replaced by coal, which is an even dirtier fuel. Underdeveloped nations that want to industrialize in order to improve their standard of living would play a major part in the increasing use of fossil fuel.

Several times the amount of carbon in the atmosphere is locked up in the *biota*, which includes all living things on the surface of the Earth, and *humus*, which is dead organic matter in the soil. The harvest of forests, the extension of agriculture, and the destruction of wetlands speed the decay of humus, which transforms into carbon dioxide and enters the atmosphere. Also, agricultural lands do not store as much carbon as the forests they replace. Forests are extensive (FIG. 14-5) and conduct more photosynthesis on the surface of the Earth than any other type of vegetation. They incorporate from 10 to 20 times more carbon per unit area than does cropland or pastureland. Forests also have the potential for storing carbon in quantities that are sufficiently large to affect the carbon dioxide content in the atmosphere.

The world is presently cutting down the forests for timber and to make way for agriculture at an alarming rate. Half of the forests in the civilized world are now gone, and the rest are being hastily

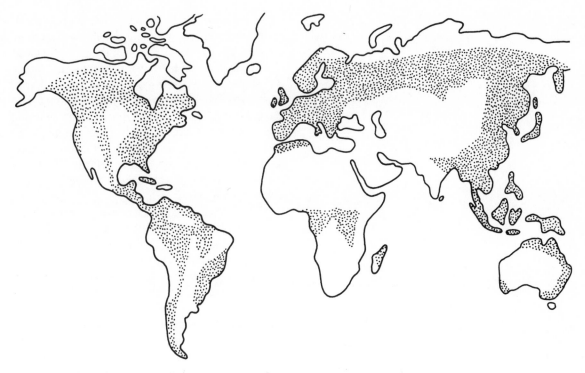

FIG. 14-5. The extent of the world's forests.

laid waste by mechanized timber harvesting and slash-and-burn agriculture. As the stores of carbon in the trees are being released into the atmosphere, the concurrent reduction of the forests is weakening their ability to remove excess carbon dioxide from the atmosphere, as well as their ability to replenish oxygen.

The ocean has by far the largest store of carbon dioxide—almost 60 times greater than the atmosphere. Carbon dioxide enters the ocean from the atmosphere by surface wave action, and the concentration of carbon dioxide in the upper 250 feet of ocean is as much as that in the entire atmosphere. In this mixed layer of the ocean, microorganisms use the carbon dioxide to make their shells, which fall to the sea bottom when the animals die. The shells eventually accumulate into thick deposits on the ocean floor, where they are converted into limestone. The limestone permanently locks up the carbon dioxide until it is subducted into the Earth. There the limestone is heated, releasing the carbon

dioxide which makes its way back into the atmosphere through volcanic eruptions.

The abyssal, by virtue of its great volume, holds the vast majority of the free carbon dioxide, and the capacity of this region of the ocean to store carbon dioxide is almost limitless. Unfortunately, carbon dioxide moves from the atmosphere into the ocean at a slow, constant rate, which is only half the rate it is being released into the atmosphere by man's activities.

SOLAR DEATH RAYS

When the atmospheric oxygen content reached a fraction of its present value some 400 million years ago, a thin veil of ozone formed between 25 and 30 miles altitude. The ozone layer blocked out most of the deadly ultraviolet light from the Sun (FIG. 14-6), and this event signaled the arrival of the land plants. The land animals followed once they were protected from one of the Sun's most harmful radiations.

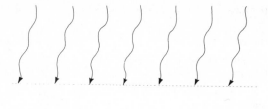

OZONE

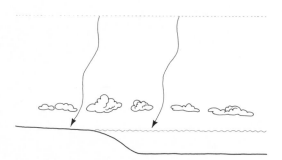

FIG. 14-6. The ozone layer.

Today, the ozone layer is being steadily depleted. Ozone-attacking chemicals such as chlorofluorocarbons—used as refrigerants, as spray can propellants, and as industrial solvents—have been irretrievably dispersed into the atmosphere, and the rate at which these chemicals are being emitted continues to grow annually. As a result, there is a measurable drop in the amount of ozone in the upper atmosphere that could lead to increases in the incidence of skin cancer among humans, have harmful effects on plants (particularly crops) and animals, and exacerbate pollution problems such as smog and acid rain.

Ozone is produced in the upper stratosphere through the absorption of solar ultraviolet radiation by oxygen molecules composed of two oxygen atoms. When the chemical bond is ruptured, ozone, an unstable molecule of three oxygen atoms, along with a single oxygen atom are produced. The ozone then decays back to an oxygen molecule and an oxygen atom. Certain chemicals released into the stratosphere compete for the free oxygen atoms and therefore interfere with ozone production.

The ozone makes up only a trace constituent of the stratosphere, with a maximum concentration of only a few parts per million of the air molecules. If the diffuse ozone layer were concentrated into a thin shell of pure ozone gas surrounding the Earth at atmospheric pressure, it would only measure about 1/8 inch thick. Furthermore, ozone destruction mechanisms are based on chain reactions, in which one pollutant molecule might destroy many thousands of ozone molecules before it is transported to the lower atmosphere where it can no longer do any harm.

Every spring for the last several years, there has been a large hole, about the size of the United States, in the ozone layer over Antarctica (FIG. 14-7). Every year, it has worsened, with about a 40 percent decrease in the amount of ozone over the past decade. The Antarctic ozone hole hovers over the South Pole and is surrounded by a ring of high ozone concentrations caused by the giant circular wind patterns around the pole, called *polar vortex*. Scientists have little doubt that a chemical process is responsible for the hole and that the chemicals are mostly man-made. There appears to be another smaller hole in the ozone layer above the Arctic that occurs between October and February. Its discovery in June 1986 came with what might be the first observable drop in atmospheric ozone of about 3 percent in a 6-year period. Depending on

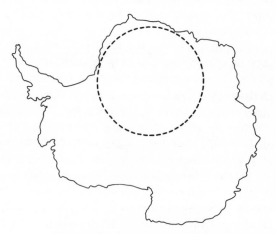

FIG. 14-7. Approximate position of the ozone hole over Antarctica.

its cause, this could be quite serious. Previously, scientists and policymakers were concerned about the consequences of the 8 percent projected decrease in ozone in the next century.

MAGNETIC FLIP-FLOPS

Geomagnetism is produced in the Earth's core by thermal, chemical, and electrical reactions. The core can be likened to a disk dynamo, which generates a magnetic field for as long as the disk is revolving. Whenever there is an imbalance resulting from either internal or external influences, the magnetic field collapses, and for unknown reasons, it is regenerated in the opposite direction. During a reversal, a compass held with the north end of the needle pointing to the North Pole would be observed to wander aimlessly for a few thousand years, and then the north end of the needle would swing around and point to the South Pole.

There have long been attempts to make correlations between changes occurring in the Earth's core, which affect the magnetic field, and events on the Earth's surface. Although changes in the magnetic field do sometimes coincide with changes in global ice volume, short-term rapid glaciations, and climatic cooling, there has yet been no conclusive proof that the relationship is anything more than coincidental.

The geomagnetic field reverses in a highly irregular manner and appears to be a random process. After several hundred thousand to millions of years, the strength of the field gradually decays over a period of 10,000 years, and then suddenly (on a geologic time scale), it reverses itself and slowly builds back to normal strength. There have been 60 complete changes of magnetic polarity over the past 20 million years, with the last reversal occurring some 700,000 years ago. Magnetic reversals have been observed in Precambrian rocks and have occurred in all subsequent geologic periods.

Evidence for magnetic reversals exists in lava flows the world over. Lava is normally iron rich, and when it cools, the iron atoms line up in the direction of the Earth's magnetic field. This forms weak magnetic fields in the rock, which can be detected by sensitive magnetometers that can determine both the direction and strength of the magnetic field. Magnetic reversals also are found in parallel bands on the ocean floor on either side of a spreading ridge, and their discovery was one means of proving the theory of seafloor spreading.

A comparison of magnetic reversals with known variations in the climate has shown in many cases a striking agreement. During a magnetic reversal when there is no magnetic field, the Earth is colder. At this time, the Earth lets down its cosmic radiation shield, the magnetosphere (FIG. 14-8). This exposes the Earth to the solar wind and cosmic radiation.

When the field first begins to weaken, cosmic rays could penetrate into the lower atmosphere and warm it. As the bombardment increases with further weakening of the magnetic field, it could influence the composition of the upper atmosphere by making more nitrogen oxides, which in turn could produce a haze that blocks out the Sun's heat and cools the Earth. The cosmic bombardment also might be detrimental to life since certain magnetic reversals coincide with the extinction of species. The magnetic field reversals also might be associated with other geological phenomena, such as intense volcanic activity, very high magnitude earthquakes, or a large meteor impact. They also could have an adverse affect on the climate. The magnetic field has weakened by about 5 percent over the past century. If this trend continues, within the next thousand years or so, the Earth's magnetic field might become very weak or nonexistent.

THE RISING SEA

Venice, Italy, is drowning because the sea is going up and the city is going down. The incidence of high tides has been increasing in frequency and magnitude since 1916. Venice begins to go underwater when the tide rises to about 2½ feet above mean sea level, which occurs about 45 times a year. The high water is destroying much of Venice's beauty by eroding the foundations of buildings and statues and by soiling paintings and frescoes. Much of the subsidence is a result of overuse of groundwater, causing the aquifer under the city to compact. The cumulative subsidence of Venice over the last 50

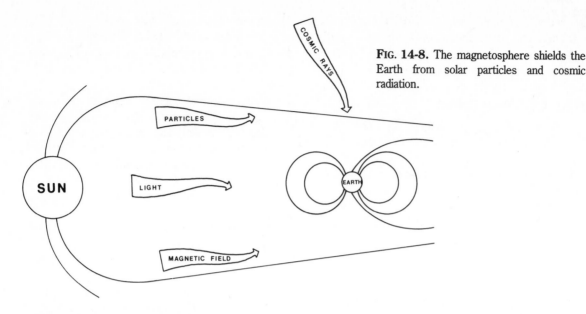

FIG. 14-8. The magnetosphere shields the Earth from solar particles and cosmic radiation.

years is slightly more than 5 inches. Meanwhile, the Adriatic Sea is rising as a result of the rise in world sea levels brought on by a global warming that is melting the ice caps (FIG. 14-9). The Adriatic has risen about 3½ inches during the twentieth century. Together with the subsidence, there is a change of more than 8 inches between Venice and the sea. This change has caused an increase in flooding from spring runoff, high tides, and storm surges.

In order to save the city, scientists are studying the feasibility of building locks to seal off from the sea the lagoon upon which Venice was built. Opponents of the project fear that eventually the lagoon would fill with silt, the canals would dry up, and Venice would no longer sit at the edge of the sea, losing much of its tourist appeal.

The ice caps cover about 7 percent of the Earth's surface area. During the last interglacial period from about 120,000 years ago to about 100,000 years ago, the climate was warmer than it is today and the melting of the ice caps caused the sea level to rise about 60 feet above what it is now.

The Arctic is a sea of pack ice whose boundary is the 50 degree Fahrenheit July isotherm. The sea ice covers and area of about 4 million square miles and its average thickness is several tens of feet. Antarctica is an ice-covered continent with an average

surface area of 6 million square miles and an average ice thickness of over 1 mile. If just the ice in the Arctic melted, it would raise the sea level by only a few feet. If both Arctic and Antarctic ice were to melt, however, the sea level would rise about 300 feet. This would move the shoreline up to 70 miles inland in most places (FIG. 14-10)—even more at low-lying deltas—and radically change the shapes of continents. Because most of the major cities of the world are either located on the coasts or along inland waterways, they would drown, with only the tallest buildings showing above the water line.

Even complete melting of the ice caps is not necessary. A warmer climate could induce an instability in the West Antarctic ice sheet, causing it to surge into the ocean. This rapid flow of ice into the sea could raise the sea level 15 feet, inundate up to 3 miles of shoreline worldwide, and flood over $1 trillion worth of real estate.

Just the opposite would happen with the onset of an ice age. At the height of the last ice age, the average global temperature was 10 degrees Fahrenheit colder than it is today, and the sea was about 300 feet below its present level. An estimated 10 million cubic miles of water were incorporated in the continental ice sheets, which covered about one-third of the land surface. The coastline of the east-

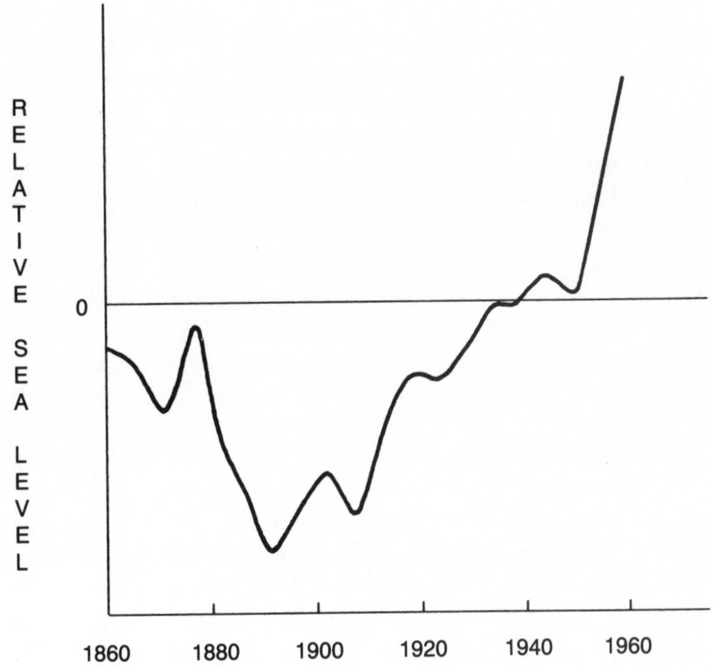

FIG. 14-9. Sea level through time.

FIG. 14-10. Parts of Europe that would be inundated if the ice caps melted.

RELATIVE SEA LEVEL

0

1860   1880   1900   1920   1940   1960

ern seaboard of the United States was about half-way out to the edge of the continental shelf, which extends eastward more than 60 miles.

On the continental shelf off the eastern United States, a step on the ocean floor that has been traced for 185 miles might represent what was once a sea cliff. Submarine canyons carved into bedrock 200 feet below sea level can be traced to rivers on land. These submerged valleys were carved by rivers when the sea level was lower during the ice age. Also, many land bridges appeared, which aided the migration of animals and man to different parts of the world.

NUCLEAR FREEZE

The past might bare some clues that indicate what could happen to the world when the sky becomes clogged with dust and smoke, as it would following a nuclear exchange. The clues are hidden in 65-million-year-old sediments that were laid down at the end of the Cretaceous period, when the dinosaurs and nearly three-quarters of all species mysteriously vanished. Only something akin to nuclear war could have destroyed life on such a large scale.

A thin layer of mud at the top of Cretaceous rocks contains crystals that show signs of being shattered by some sort of shock mechanism. The mud contains a high concentration of iridium, a member of the platinum group, which is very rare on Earth but is relatively abundant in meteors. The mud also shows a thin layer of common soot. Instead of nuclear missiles, a massive shower of meteors might have rained down on the Earth, injecting huge quantities of dust into the atmosphere and setting global forest fires that produced a large blanket of smoke. The dust and the smoke would have effectively blocked out the Sun, causing the Earth to cool dramatically in a very short time.

The nuclear arsenals of the world presently contain the explosive equivalent of 4 tons of TNT for every man, woman, and child on Earth. This reflects an enormous overkill factor, considering that 1 pound of TNT is more than enough to kill a person.

Half the human population of the world, mostly in the Northern Hemisphere, can be expected to die within the first few weeks following a full-scale nuclear war. In addition to tremendous blast waves that can knock down substantial buildings 5 miles downrange from ground zero, nuclear weapons produce vast quantities of highly radioactive fallout that is dispersed the world over, and even those people living in the Southern Hemisphere would not be immune from its effects. People living in major cities or near military targets would be inflicted with radiation burns, as well as burns from the gigantic fireball and the fires it sets. Burning cities also would give off dangerous toxic fumes, which only makes survival that much more difficult. In addition to cities, forests near targeted areas would catch fire from thermal radiation and would burn out of control.

Past scenarios for nuclear war only considered the damage to industry and tallied up human deaths, but failed to take into account the ecological damage to the Earth. Disease from decomposing, unburied corpses, both humans and animal, would spread like wildfire. Some plants, including crops, have a low tolerance for radiation, and many animals have a lower tolerance than do humans. Pests, like insects and rats, have a considerably higher tolerance for radiation, and without natural predators, they would multiply rapidly, spreading disease in epidemic proportions.

The soil and water would be heavily contaminated with radiation and toxins. Sulfur and nitrogen compounds released into the air from burning cities would rain out as destructive acids. Large numbers of nuclear detonations also would destroy the ozone layer, leaving any surviving inhabitants on the surface exposed to dangerous levels of ultraviolet radiation.

The total smoke emission from a full-scale nuclear war could be as much as 300 million tons. If it were spread evenly around the world, the smoke could reduce the amount of sunlight reaching the ground by as much as 95 percent. Typically, heavier smoke would cover the target zones where noon would be as dark as a moonlit night. The detonation of large numbers of nuclear weapons near the ground also would inject several hundred million tons of soil and dust into the atmosphere.

The reduction of sunlight by the smoke and dust

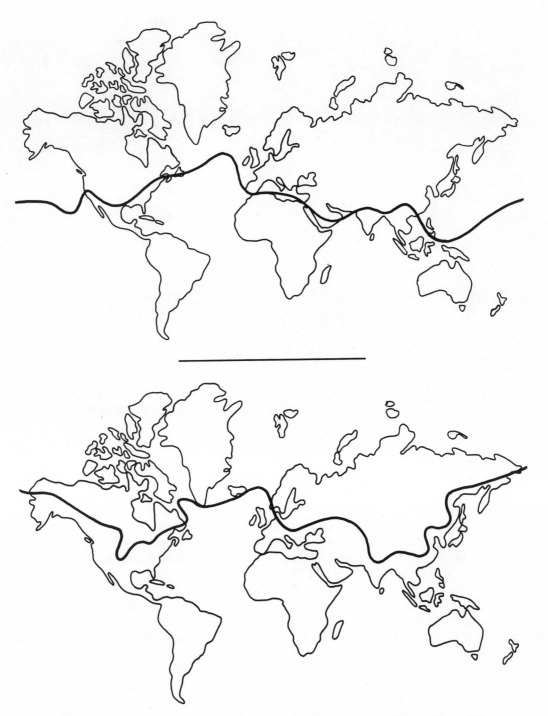

FIG. 14-11. The July smoke line (top), above which sunlight strength is equal to twilight or darker during daytime 3 weeks following a general nuclear war. The July freeze line (bottom), above which temperatures are freezing or below 3 weeks following general nuclear war.

FIG. 14-12. Detonation of a hydrogen bomb
at Bikini Atoll. Altitude of 12,000; distance of 50 miles northwest of target.

(Courtesy of U.S. Navy)

could cool the Earth by several tens of degrees Fahrenheit (FIG. 14-11), and the cooling could persist from several months to over a year. This could bring freezing conditions to temperate zones in the middle of summer. Another consequence is that plants, which depend upon warm temperatures and sunlight for their growth, would wither and die. Therefore, any attempt to grow crops in order to produce enough food to survive would be useless.

On the ocean there would be increased ice floes, but as a whole, the ocean would not freeze because of its high heat capacity and high thermal inertia. The large temperature difference between the land and the sea would produce violent coastal storms. The runoff would carry huge amounts of radioactive poi-

sons, toxic chemicals from destroyed cities, and disease-ridden organic matter into the ocean. In effect, the ocean would become one gigantic nuclear and chemical dumping ground. The continual darkness and poisons would kill the primary producers, such as phytoplankton, with tragic consequences on up the food chain. As a result, the seas could no longer be counted on to supply human dietary needs. In tropical areas where indigenous organisms are particularly temperature sensitive, vast numbers of marine species quickly could become extinct. The ocean would experience a cataclysmic die out of species on a scale similar to the largest extinctions in the Earth's long history—only this time it would be caused by the hand of man (FIG. 14-12).

15

The Pollution Solution

MAN is the most highly evolved creature on the Earth and has long regarded the planet as his for the taking. Unfortunately, the environmental impact of this attitude has been profound and often destructive. Earlier, when man was a hunter-gatherer, he needed to be highly attuned to the world around him in order to take advantage of its resources and avoid its dangers. Now, modern society lives in an artificial world, far removed from the balance of nature.

The industrialization that made this modern lifestyle possible is in the process of polluting the very environment in which man and the rest of the world are required to live. The release of cancer-causing chemicals and other hazardous substances into the atmosphere is far greater and more widespread than it was ever before suspected. On a grand scale, acid rain is destroying forests in the Unites States and Canada and killing fish in Sweden. Dangerous chemicals seep into the groundwater supply, placing many municipalities at risk from contaminated drinking water. The dumping of toxic wastes into the ocean might cause irreversible changes in aquatic ecosystems. The disposal of the ever-mounting stockpiles of nuclear wastes is highly crucial because they remain a hazard to life for thousands of years. If the solutions to these problems and others are not forthcoming, man might find that the Earth is no longer a healthy place to live.

DANGER IN THE AIR

Air pollution has become a growing threat to health and welfare because of the ever-increasing emissions of air contaminants into the atmosphere. Each day, an average adult takes in about 30 pounds of air, as compared to 2.8 pounds of food and 4.5 pounds of water. Therefore, the cleanliness of the air should be just as important, if not more so, as the cleanliness of food and water. There are also competing natural pollutants in the air, including salt particles from ocean spray and breaking waves, pollen and spores released by plants, smoke from brush and forest fires, and wind-blown and volcanic dust. In the Great Smoky Mountains of eastern Tennes-

FIG. 15-1. Approximately 20 percent of pollution is removed by rain.
The remaining 80 percent is removed through dry deposition.

see, there exists a natural photochemical smog similar to that found in big cities. Whenever hydrocarbons, in this case pine sap, react with sunlight and the damp surface air, they produce a smoky haze, and that is why the mountains were given their name.

Atmospheric pollutants are grouped into two jor categories. *Primary pollutants* are those emitted from identifiable primary sources, such as smoke stacks and automobile exhausts. The transportation sector is the most significant and accounts for more than half of the primary air pollution. *Secondary pollutants* are produced in the atmosphere by chemical reactions taking place among the primary pollutants.

Many reactions that produce secondary pollutants are triggered by sunlight and are called *photochemical reactions.* Nitrogen oxides produced in factory furnaces and by automobiles absorb solar radiation and initiate a chain of complex chemical reactions. In the presence of organic compounds, these reactions result in the formation of a number of undesirable secondary products that are very unstable, irritating, and toxic. One such substance is ozone, and although it is beneficial in the upper atmosphere, where it screens out harmful ultraviolet light from the Sun, it is poisonous near the ground.

Weather plays a vital role in distributing, and thereby diluting, air pollution in the atmosphere. When bad pollution days occur, they are not necessarily the result of increased contamination, but are more likely because of changes in atmospheric conditions. These conditions include the strength of the wind and the stability of the air, which determine how rapidly pollutants are diluted by mixing with the surrounding air after leaving the source (FIG. 15-1). Because of the direct effect of the wind speed, the concentration of pollutants is half as much with a 10-mile-per-hour wind as it is with a 5-mile-per-hour wind. Also, the stronger the wind, the more turbulent is the air. Therefore, strong winds mix polluted air with the surrounding clean air more rapidly.

The vertical motion of the air determines to what extent polluted air mixes with cleaner air above. When the air is unstable, smoke and exhaust fumes are carried upward by air currents, mixed with the upper air, and dispersed by winds aloft. If the air is stable, however, updrafts are suppressed and the overlying warm air acts like a lid, holding the pollution in.

Thousands of tons of dangerous chemicals are released into the air by chemical companies around the country. One of these substances is methyl isocyanate which is used in making pesticides. An industrial disaster involving this substance caused the deaths of 2000 people in Bhopal, India, in late 1984.

Over 200 hazardous chemicals are vented into the air, and a few industrial plants release large quantities of known carcinogens, or cancer-causing agents. Although there is yet no conclusive evidence concern-

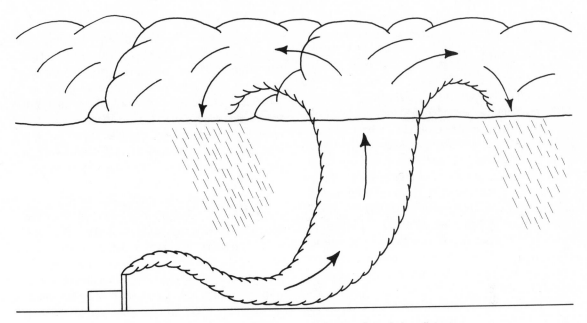

FIG. 15-2. Acid rain production by cloud scavaging of air pollutants.

ing the health effects of these toxic pollutants on the population, many scientists are concerned with the long-term exposure to air contaminants. Many of these substances are rained out of the atmosphere and end up in soils, rivers, lakes, and seas, where they are concentrated by chemical and biological agents.

In addition to the poisons dumped into the atmosphere in alarming amounts, man-made particulates and aerosols account for an estimated 15 million tons of soot and dust suspended in the atmosphere at any one time. All these unnatural substances in the atmosphere might produce dramatic changes in the climate, resulting in a greenhouse effect. The increase in global temperatures caused by the greenhouse effect might in turn cause a cataclysmic melting of the polar ice caps.

ACID RAIN ON THE PLAIN

One of the results of burning tremendous quantities of coal and oil is that millions of tons of sulfur dioxide and nitrogen oxides are discharged into the atmosphere each year. These gases combine with oxygen and atmospheric moisture to produce sul-

furic and nitric acids, which then fall as rain or snow (FIG. 15-2). Measurements of the acidity of rain and snow reveal that in parts of the eastern United States, eastern Canada, and northwestern Europe precipitation has changed from a nearly neutral solution at the beginning of the Industrial Revolution two centuries ago to a dilute solution of sulfuric and nitric acids today. In the most extreme cases, the rain had the acidity of vinegar.

Current research efforts in North America and Europe have intensified and are directed toward determining both the direct and indirect effects of increased environmental acidification on various aquatic and terrestrial species. Acidified lakes in New York's Adirondack Mountains are totally devoid of fish. In Canada, 14,000 lakes are threatened by acid rain. In Sweden, it is estimated that more than 15,000 lakes are without fish because of acid rain.

Because many pollutants can remain in the atmosphere for days, they tend to travel long distances from their sources and even cross international borders, causing a host of political problems. Much of the acidity found in the eastern United States as well as in Canada originated hundreds of miles away in the industrialized regions of the Midwest and the

Ohio Valley. Sweden and Norway are not heavy pol-luters, and yet they are constantly being bombarded by acid rain produced by the heavily industrialized areas of Great Britain and West Germany.

Acid rain is possibly the most studied and the least acted upon of any type of pollution. The source of the problem has been identified for decades, leav-ing little doubt about the basic chemistry that turns industrial and automobile emissions into acid rain. However, governments are slow to legislate man-datory emission controls. For one thing, scrubbers installed on smokestacks to remove sulfur dioxide are expensive, and companies might lose their com-petitive edge by being forced to make such an in-vestment. The burning of low-sulfur coal is another alternative, but this solution would mean an increase in fuel costs that ultimately would be passed on to the consumer.

Unlike most atmospheric pollutants, acid rain has not yet been shown to have any direct adverse effects on human health, but its damaging effect on the environment is proven beyond a doubt, however. Acid rain might have a toxic effect on many plants and animals by changing the chemistry of their en-vironment. Streams and lakes in many parts of the world have become so acidic from acid-rain runoff that fish populations have been virtually wiped out. Fish, in particular, have a low tolerance for high acidity levels, and a drop in fish populations can trigger changes in other organisms as well. Some soils have become so acidic they no longer can grow crops. The acid rain percolates into the soil, damaging the fine root systems of the plants and leaching valuable nutrients from the soil. Acid rain also produces ad-verse effects when it falls on foliage. It is resulting in the destruction of the great forests of North America, Europe, China, and Brazil.

Metals have been leached out of geologic for-mations or conducting pipes made of copper or alu-minum. Metal structures such as bridges, equipment, and even the Statue of Liberty have be-come corroded and weakened. Marble buildings and statues have been defaced by acid rain, which dis-solves marble, destroying much of their artistic beauty.

CREEPING POISON UNDERGROUND

Throughout the United States, dangerous chemicals are leaching out of as many as 16,000 land-fills; pesticides and fertilizers are penetrating the ground; and the contaminants are percolating down through layers of soil into vast underground water reservoirs called *aquifers*. Subsurface water, which underlies millions of square miles throughout the country and encompasses an estimated 65 quadril-lion (65 followed by 15 zeros) gallons of water, is becoming increasingly polluted.

Man-made organic chemicals, heavy metals, pesticides, and other toxic substances are seeping into the ground from landfills, buried gasoline tanks, septic systems, radioactive waste sites, farms, mines, and a host of other sources. They might con-taminate the nation's aquifers to such an extent that in the coming years up to a quarter of the nation's groundwater might be unusable. The groundwater problem has become the primary environmental challenge of the century, and cleaning up the con-tamination will be expensive, difficult, and in some cases, impossible.

Many thousands of wells across the nation have been contaminated by highly concentrated chemicals that spread through aquifers and exceed the federal limits for safe drinking water. In California's Silicon Valley, solvents used in the manufacture of computer chips have leaked from buried storage tanks into the water supplies of several communities. A New Jer-sey landfill that contained some 9 million gallons of hazardous wastes contaminated the aquifer and forced the shutdown of water wells in Atlantic City. In Florida alone, there are 6000 lagoons and ponds filled with toxic waste, and contaminants were found in the groundwater, forcing the shutdown of more than 1000 wells. Although it has yet to be proven unequivocally that small doses of toxic chemicals are harmful to man, federal regulators have proceeded on the assumption that long-term exposure to even minute amounts of most organic chemicals en-dangers the public's health.

At least 1 to 2 percent, and possibly as much as 10 percent, of the groundwater supply across the country has already been contaminated. However, the contamination might become even greater as slow-moving patches of pollutants advance on populated areas, typically at a rate of about 1 foot per day. The increase of groundwater use (12 percent since 1980) also will make matters worse because pumping wells speed the flow of water in aquifers and thus, the flow of contaminants.

More than 50 percent of Americans depend on groundwater for drinking and irrigation (FIG. 15-3), and in 1985, Americans pumped 100 billion gallons of water per day from aquifers. Most of the drinking water comes from wells that are less than 300 feet deep. Many central and western states draw more than half their water from the ground. Therefore, contamination in these areas could prove to be catastrophic because there is not enough surface water to satisfy domestic, industrial, and agricultural demands. In several million rural households, aquifers are the only source of water.

In many instances when their aquifers become contaminated, people complain that their well water smells or tastes bad. In other instances, routine monitoring of industrial waste lagoons or landfills reveal that the chemicals are not being contained, which prompts the testing of nearby wells (FIG. 15-4). Once the pollution is found, determining its extent is difficult. Pools of chemicals generally advance at the rate of the subterranean flow, flatten-ing and elongating as they go. The chemicals might react with the porous rocks of the aquifer or become chemically bound to them. In the absence of air, they might break down differently than they would on the surface. In order to track down the pollution, scientists must determine the direction and speed of groundwater flow; the location, thickness, and composition of the aquifer; and its ability to transmit and store water. In order to determine these things, they must drill a series of water wells across a suspected area—a procedure that is expensive and time-consuming.

Cleaning up an aquifer is nearly an impossible task. In the simplest cases, when the contamination is fairly localized, the pollutants can be pumped out through wells. When the chemicals have spread over a wide area, they can be blocked or encircled with impermeable clay. Even then, however, these methods only work when the contamination comes from a single source and covers a limited area. If the contamination is irreversible, the only recourse is to treat the water at the well head.

Any contaminated water can be rendered drinkable, but it is very expensive to clean up large volumes of water. Cleaning up the landfills and waste lagoons is also costly, and most states, which bear the burden of monitoring and safeguarding groundwater, cannot afford the enormous investment with-

FIG. 15-3. Groundwater irrigation in Kansas.

out federal government support. Ultimately, improved methods of dealing with waste problems and better knowledge of the underground environment will help solve the problems in the future. Unfortunately for much of the nation's groundwater, however, past mistakes might have already made recovery too late.

THE OCEAN CESSPOOL

Along the eastern coast of the Americas, the sea lies on a wide and gently sloping continental shelf that extends over 60 miles, reaching a maximum depth of 600 feet at the shelf edge. In contrast, along the western coast, the water descends rapidly to great depths a short distance offshore. The sea off the eastern coast is dominated by freshwater discharges from the great coastal plain estuaries from the St. Lawrence River in the north to the great Amazon River in the tropics. In marked contrast, the western coast is dominated by periodic upwelling of bottom water from along the Oregon coast to the coast of Peru.

The coastal seas of the world are among the most fragile and sensitive environments. Some laws passed to regulate the wastes that can be dumped into the oceans were based on inadequate knowledge of the sea, and compliance with those laws

might do little to make the ocean cleaner. Some of the changes that human activity have wrought in the ocean environment are irreversible, such as the damming of rivers, which make their discharges smaller, and the building of ports at the mouths of estuaries, which changes the patterns of flow and alters the coastal habitat.

Offshore oil spills are perhaps the most damaging of all coastal pollution. In January 1969, a major oil-well blowout at Santa Barbara, California, released some 10 trillion gallons of crude oil into the ocean during the first 100 days, with immediate and disastrous local ecological consequences. This dramatic event brought about public awareness of the enormously high level of industrial pollution in both the atmosphere and the waters, which led Congress to enact the National Environmental Policy Act of 1969 and later, the 1972 Federal Water Pollution Control Act Amendments. The regulations required anyone who discharges wastes into public waters to obtain a permit that outlined limitations and monitoring requirements for the discharge.

Major oil spills continued to occur, however, and the *Argo Merchant* spill off Cape Cod in 1976 (FIG. 15-5) made it necessary to reexamine oil-spill response strategies, especially in productive fishing grounds. Another major oil-well blowout in the Gulf

(Courtesy of USGS)

FIG. 15-4. Testing water for contamination.

of Mexico in June 1979 and its fouling of Texas beaches continued to dramatize the dangers of offshore drilling. Massive and continuous spilling of oil in the Persian Gulf as a direct result of the Iran-Iraq war threatens to endanger the entire ecology of the Persian Gulf area, and could have serious implications for decades to come.

A number of environmental processes determine the outcome of an oil spill once it has occurred. The overall national response strategy for major oil spills include the use of computer models to predict the course of the spill. The principal use of these models is oil-spill contingency planning. The transport of pollutants is affected by the ocean currents, the local wind, and the tides. An active response might result in a mechanical cleanup (FIG. 15-6) or containment, or the use of detergents or other chemicals to breakup the oil spill. Unfortunately, such chemicals might cause additional toxic effects on the ecology of the region. The success of the operation ultimately depends on accurate weather forecasts, particularly the wind speed and direction.

There is a need for additional research into the dispersion, spreading, and subsurface transport of oil spills. This information is critical in order to determine the extent of the oil spill as well as the fate of the coastal ecology.

Although oil spills are highly visible and attract a lot of public attention, the dumping of toxic wastes into the ocean is an insidious and potentially more serious problem. Approximately 8 million tons of toxic wastes are dumped into rivers and coastal waters each year (FIG. 15-7). The long list of toxic substances includes chemicals such as chlorinated hydrocarbons, benzenes, trichloroethylene, toluene, polychlorinated biphynols (PCB's), dioxin, solvents, organic chemicals, pesticides, and fertilizers, as well as fibrous asbestos and heavy metals. Some of these toxic pollutants are powerful carcinogens and mutagens, and many are nonbiodegradable, persisting in the environment for long periods. Even very small amounts of these pollutants below detectable limits are dangerous. What is even worse though, is that the toxic pollutant tends to become concentrated to

(Courtesy of NOAA)

FIG. 15-5. Oil spill from the *Argo Merchant* in December 1976.

FIG. 15-6. Oil spill cleanup on the beaches of Sandy Hook, New Jersey from a collision between a barge and oil tanker.

FIG. 15-7. Water pollution in the Cumberland River north of Nashville, Tennessee.

lethal levels as it progresses up the food chain. At the base of the marine food chain are phytoplankton, which act as concentrators of a pollutant. These are eaten by zooplankton, which in turn are fed upon by crustaceans and fish, which are ultimately eaten by man.

The cost of land disposal of toxic waste is escalating, forcing coastal metropolitan areas to dump industrial waste and raw sewage directly into the ocean. Federal funds are not available to meet the over $100 trillion needed for the nation's cities to dispose of their waste; therefore, the pressure on ocean dumping is mounting. For a typical ocean dumping site along the East Coast, barge loads of waste are taken out to about 100 miles and dumped beyond the continental shelf. After a day or so, the pollutants are diluted by a factor of about 50,000, but there remains a question whether even a dumping dilution of 100,000:1 is sufficient. The pollutants tend to concentrate in regions where the seawater density changes, such as an oceanic thermocline or an ocean front. These frontal areas are also the feeding grounds for fish, however. The meandering currents of the Gulf Stream, laden with fish and other ocean life, actually sweep over the dump sites.

Of all the alternative methods of getting rid of hazardous waste, none has engendered as much controversy as the burning of toxic substances at sea. One plan involves the use of an incinerator ship to burn toxic wastes off New Jersey, and would be only a stopgap measure to treat toxic liquids until better methods to reduce or recycle waste could be developed. Ocean incineration would only be suitable for treating 5 to 8 percent of all hazardous waste, but the chemicals that could be destroyed by the technology are among the most toxic. Incineration at sea is one of the few methods available to detoxify hazardous waste that is highly chlorinated. There are many unresolved questions concerning the potential risks to health and the environment. Better methods are needed to measure whether dangerous compounds have actually been destroyed by burning, and to identify what compounds are being emitted into the atmosphere and ultimately end up in the ocean.

A WATERY GRAVE FOR NUCLEAR WASTES

The disposal of radioactive wastes has received much attention in recent years because of increasing concerns of environmental effects over very long periods and because of the expansion of nuclear technology throughout the world (FIG. 15-8). Much of the effort to find disposal sites has been expended on exploring continental geologic formations, but two-thirds of the planet's geology is submarine geology, which has been largely ignored.

High-level nuclear waste is the most difficult radioactive waste to dispose of because of its radiation output, heat, and longevity. One idea is to process and package the nuclear wastes and bury them in deep, stable geologic formations under the seabed. One such formation is a dark, chocolate brown, fine-grained, extremely cohesive sediment called *abyssal red clay*. It is generally found in deep open-ocean areas seaward of continental margins and abyssal plains. The clays are produced in the ocean by a slow, steady rain of windblown particles from land, meteoric and volcanic dust, insoluble organic debris, fine precipitates from midocean ridge hydrothermal systems, and authigenic (formed in place) minerals, all of which accumulate on the ocean floor. In the absence of redistribution by bottom currents or turbidity currents, these materials produce a unique geologic formation that blankets about 30 percent of the seafloor.

Some of these deep-sea formations might be able to contain nuclear wastes because they have desirable barrier properties and stable geologic histories. If a typical isolation period of 1 million years is required, then the formation should exhibit a stable geologic record at least ten times that long. The best locations are away from lithologic plate boundaries where geological disturbances such as large earthquakes, volcanoes, or faults disrupt the uniform layers of sediment. The sites also must be located away from areas of steep or rugged topography, where slumping or erosion by deep-sea currents might disrupt the continuous deposition of fine sediment. Also, areas that contain natural resources such as fisheries or petroleum and mineral deposits should be avoided.

The canisters containing the radioactive waste could be placed in holes drilled into the bottom sediments several hundred feet deep or deeper. The emplacement hole would then be closed, forming an effective barrier to seawater. After emplacement, the location of each canister in the sediment could be precisely determined by an acoustic-magnetic sensor system.

It is generally thought that the best place to store nuclear waste is underground. Salt domes and granite make the most stable geologic formations on the continents. However, mine repositories are very expensive and require additional costs for backfilling and shaft sealing. Also, there is no absolute guarantee that, after the containers age and begin to leak, they will not contaminate the nearby groundwater system. Therefore, reliable predictions concerning the possible migration of radioactive fluids through geologic formations surrounding the repository must be made. The formation must remain stable for 1 million years or so without earthquakes or other tectonic activity.

Finally, other methods of disposal also should be looked into. The cost of depositing radioactive wastes on the continents or under the ocean floor is comparable. The advantage of using the ocean floor is that there is no danger of contaminating the ocean of the groundwater supply, or of tampering by future generations of mankind.

CLEANING UP OUR ACT

The world-famous French oceanographer Jacques Cousteau indicated in 1985 that the world's oceans were suffering from a dangerous decrease in vitality. After diving extensively, and taking hundreds of measurements and hundreds of sam-

(Courtesy of U.S. Department of Energy)

FIG. 15-8. The Three Mile Island nuclear power plant, Harrisburg, Pennsylvania.

ples of water and sediment, he realized that the drop in animals was much larger than could be explained by chemical pollution alone. The decline was also a result of what Cousteau called *mechanical destructions*, such as dynamite fishing, fishing in spawning grounds, using fine-mesh fishing nets that take the young as well as the adults, diverting rivers, filling in marshes, and doing other destructive activities. If man's attitude of exploiting the oceans for short-term gains is not changed and if the consequences of this exploitation are not made known to everyone, the world faces a catastrophe in the long term.

Nor is one ocean more or less healthy than another because water moves, making any differences temporary. This is why pollutants like DDT were found in the livers of penguins in the Antarctic where there is no pollution. In 90 years, there will not be a single drop of water in the Mediterranean Sea that is there today. The pollutants in that vast cesspool will finally come to pollute the rest of the oceans. The same is true for the Caribbean, the North Sea, the Gulf of Finland, and other polluted seas. Although rivers and enclosed or semienclosed seas are presently in worse shape than the open ocean, that might not be true 10 to 20 years from now.

On November 1, 1986, a fire at a Swiss chemical factory in Basel, containing nearly 1400 tons of fungicides, pesticides, and other agricultural chemicals, caused one of Europe's worse pollution incidents of the past decade. Over 30 tons of toxic chemicals were spilled into the Rhine River when firefighters attempted to put out the fire. On one stretch of the upper Rhine from Basel, Switzerland, to Karlsruhe, West Germany, over 100 miles to the north, the Rhine was nearly totally devoid of life. At spots along the river, a purple sludge coated the banks. More than 0.5 million fish and eels are thought to have died as a 25-mile-long chemical slick moved down the river, causing ecological damage that scientists estimate will take 10 years to repair. At many spots along the 820-mile river, supplies of drinking water had to be cut off, and floodgates were closed to protect tributaries from contamination. All along the Rhine, which ultimately empties into the already highly polluted North Sea, is one of Western Europe's most heavily industrialized regions.

The accident substantially set back efforts to clean up the Rhine that had been going on for more than a decade.

Scientists in Sweden and Denmark warned in 1986 that the world is on the brink of a new form of environmental catastrophe. The stretch of water between Sweden and Denmark, called the Kattegat Sea, is rapidly dying. It is so polluted and starved of oxygen that it is fast losing the capacity to support sea life. The first indications that the Kattegat was endangered came in the 1960s, when lobsters disappeared from the southern part of the sea. More recently, thousands of dead fish were washed up on beaches at Kielfjorden, West Germany, close to the Danish border, and fishermen reported large hauls of dead fish. The Swedish Fisheries research vessel *Ancylus* found a severe lack of oxygen at all depths below 75 feet, and fish samples showed that the situation was extremely serious.

In addition to pollution from Denmark and southern Sweden—mainly pesticides and fertilizers draining from farm land—the Kattegat is a difficult sea from an ecological point of view because that is where the waters of the Baltic meet the much saltier currents of the North Sea. Danish biologists estimate that around 30 million fish have already died in the Kattegat. This is an even greater tragedy because the bulk of the Scandinavian diet is fish.

In many parts of the world, environmental protection is forced to take a back seat to other concerns, including the loss of jobs, bankrupt businesses, decreased productivity, and a host of other economic problems. Many in America feel that a clean and healthy environment is a luxury the nation can ill afford. It takes money to clean up the environment, and pursuing this goal could damage the economy. It provides little solace to unemployed workers or bankrupt businesses when they are told that they must pay the price for clean air and water. The ongoing opinion is that the environment eventually will be cleaned up sometime. This means that the burden must be placed on the shoulders of future generations who will ultimately have to pay for the selfish, short-term economic gains made by this generation. It would be a great wonder if they do not hate us for it.

Bibliography

The following references are provided for further reading.

ORIGIN OF LAND AND SKY

Anderson, Don L. "The Earth as a Planet: Paradigms and Paradoxes." *Science* 223 (January 27, 1984): 347–354.

Begley, Sharon. "A New Window on the World." *Newsweek* (December 22, 1986): 56.

Boss, Alan P. "The Origin of the Moon." *Science* 231 (January 24, 1986): 341–345.

Herbst, William and George E. Assousa. "Supernovas and Star Formation." *Scientific American* 241 (August 1979).

Jacobs, J.A. *The Earth's Core*. San Diego, Cal: Academic Press, Inc. 1975.

Kerr, Richard A. "Where Was the Moon Eons Ago?" *Science* 221 (September 16, 1983): 1166.

Lewis, John S. "The Chemistry of the Solar System." *Scientific American* 230 (March 1974): 51–65.

Maddox, John. "Origin of Solar System Redefined." *Nature* 308 (March 15, 1984): 223.

"Making the Moon from a Big Splash." *Science* 226 (November 30, 1984): 1060.

O'Nions, R.K., P.J. Hamilton, and Norman M. Evensen. "The Chemical Evolution of the Earth's Mantle." *Scientific American* 242 (May 1980): 120–133.

"Opening Doors to the Core, and More." *Science News* 131 (January 3, 1987): 9.

Schramm, David N., and Robert N. Clayton. "Did a Supernova Trigger the Formation of the Solar System?" *Scientific American* 239 (October 1978): 124–139.

Wetherill George W. "The Formation of the Earth from Planetesimals." *Scientific American* 224 (June 1981): 163–174.

THE SUPER OCEAN

Boucot, A.J., and Jane Gray. "A Paleozoic Pangaea." *Science* 222 (November 11, 1983): 571–580.

Gambles, Peter. "Death of an Ancient Ocean." *Nature* 312 (November 29, 1984): 400–401.

Holland, H.D., B. Lazar, and M. McCaffrey. "Evolution of the Atmosphere and Oceans." *Nature* 320 (March 6, 1986): 27–33.

Ingersol, Andrew P. "The Atmosphere." *Scientific American* 249 (September 1983): 162–174.

Kerr, Richard A. "Plate Tectonics Goes Back 2 Billion Years." *Science* 230 (December 20, 1985): 1364–1367.

Leg 89 Staff of the Deep-Sea Drilling Project. "The Mesozoic Superocean." *Nature* 302 (March 31, 1983): 381.

Rossow, William B., Ann Henderson-Sellers, and Stephen K. Weinreich. "Cloud Feedback: A Stabilizing Effect for the Early Earth?" *Science* 217 (September 24, 1982): 1245–1247.

Toon, Owen B. and Steve Olson. "The Warm Earth." *Science 85* (October 1985): 50–57.

THE GREAT CHEMICAL FACTORY

Allman, William F., ed. "Life With Light." *Science 86* 7 (June 1986): 9.

Brock, Thomas D. "Precambrian Evolution." *Nature* 288 (November 20, 1980): 214–215.

Cloud, David. "The Biosphere." *Scientific American* 249 (September 1983): 176–189.

Ford, Trevor D. "Life in the Precambrian." *Nature* 285 (May 22, 1980): 193–194.

Gilmore, V. Elaine. "Did Life Begin in Clay?" *Popular Science* 227 (September 1985): 32.

Groves, David I., John S. R. Dunlop, and Roger Buick. "An Early Habitat of Life." *Scientific American* 245 (October 1981): 64–73.

Lewin, Roger. "Computers Track the Path of Plant Evolution." *Science* 219 (March 11, 1983): 1203–1205.

————. "Extinction and the History of Life." *Science* 221 (September 2, 1983): 935–937.

"Life's First Building Block: Made of Clay?" *Newsweek* (April 15, 1985): 100.

Macdonald, Ken C., and Bruce P. Luyendyk. "The Crest of the East Pacific Rise." *Scientific American* 224 (May 1981): 100–116.

Valentine, James W., and Eldridge M. Moores. "Plate Tectonics and the History of Life in the Oceans." *Scientific American* 230 (April 1974): 80–89.

THE GLOBAL GARBAGE DISPOSAL

Bonatti Enrico, and Kathleen Cane. "Oceanic Fracture Zones." *Scientific American* 250 (May 1974): 40–51.

Broeker, Wallace S. "The Ocean." *Scientific American* 249 (September 1983): 146–160.

Courtillot, Vincent, and Gregory E. Vink. "How Continents Break Up." *Scientific American* 249 (July 1983): 43–49.

Francheteau, Jean. "The Oceanic Crust." *Scientific American* 249 (September 1983): 114–129.

Kerr, Richard A. "Sea-Floor Spreading Is Not So Variable." *Science* 223 (February 3, 1984): 472–473.

Lewis, Brian T.R. "The Process of Formation of Ocean Crust." *Science* 220 (April 8, 1983): 151–156.

Mutter, John C. "Seismic Images of Plate Boundaries." *Scientific American* 254 (February 1986): 66–75.

Sclater, John G., and Christopher Tapscott. "The History of the Atlantic." *Scientific American* 240 (June 1979): 156–174.

Siever, Raymond. "The Steady State of the Earth's Crust, Atmosphere and Oceans." *Scientific American* 249 (June 1974): 72–79.

Weisburd, Stefi. "Drilling Discoveries in the Pacific." *Science News* 131 (February 14, 1987): 102–103.

VOLCANOES OF THE DEEP

Burke, Kevin C., and J. Tuzo Wilson. "Hot Spots on the Earth's Surface." *Scientific American* 235 (August 1976): 46–57.

Downey, W.S., and D.H. Tarling. "Archaeomagnetic Dating of Santorini Volcanic Eruptions and Fired Destruction Levels of Late Minoan Civilization." *Nature* 309 (June 7, 1984): 519–523.

Hekinin, Roger. "Undersea Volcanoes." *Scientific American* 251 (July 1984): 46–55.

LaMarche, Valmore C., Jr., and Katherine K. Hirschboek. "Frost Rings in Trees as Records of Major Volcanic Eruptions." *Nature* 307 (January 12, 1984): 121–126.

Strommel, Henry, and Elizabeth Strommel. "The Year Without a Summer." *Scientific American* 240 (June 1979): 176–186.

Strothers, Richard B. "The Great Tambora Eruption in 1815 and Its Aftermath." *Science* 224 (June 15, 1984): 1191–1197.

Strothers, Richard B., and Michael R. Rampino. "Historic Volcanism, European Dry Fogs, and Greenland Acid Precipitation, 1500 B.C. to A.D. 1500." *Science* 222 (October 28, 1983): 411–412.

Vink, Gregory E., W. Jason Morgan, and Roger R. Vogt. "The Earth's Hot Spots." *Scientific American* 252 (April 1985): 50–57.

THE WORLD OF ICE

Bailey, Ronald H. *Glacier.* Alexandria, Va.: Time-Life Books, 1982.

Bowen, D.Q. "Antarctic ice surges and theories of glaciation." *Nature* 283 (February 14, 1980): 619–620.

Covey, Curt. "The Earth's Orbit and the Ice Ages." *Scientific American* 250 (February 1984): 58–66.

Fodor, R.V. "Explaining the Ice Ages." *Weatherwise* (June 1982): 109–114.

Kerr, Richard. "An Early Glacial Two-Step?" *Science* 221 (July 8, 1983): 143–144.

_____. "Ice Cap of 30 Million Years Ago Detected." *Science* 224 (April 13, 1984): 141–142.

_____. "The Sun Is Fading." *Science* 231 (January 24, 1986): 339.

Matthews, Samuel, W. "Ice On the World." *National Geographic* (January 1987): 79–103.

Rodak, Uwe. "The Antarctic Ice." *Scientific American* 253 (August 1985): 98–106.

Williams, George E. "The Solar Cycle in Precambrian Time." *Scientific American* 255 (August 1986): 88–96.

RIVERS IN THE SEA

Barber, Richard T., and Francisco P. Chavez. "Biological Consequences of El Niño." *Science* 222 (December 16, 1983): 1203–1210.

Friend, P.F. "Storms in the Abyss." *Nature* 309 (May 17, 1984): 212.

Hollister, Charles D., Arthur R.M. Nowell, and Peter A. Jumars. "The Dynamic Abyss." *Scientific American* 250 (March 1984): 42–53.

Kerr, Richard A. "Are the Ocean's Deserts Blooming?" *Science* 220 (April 22, 1983): 397–398.

_____. "Small Eddies Are Mixing the Oceans." *Science* 230 (November 15, 1985): 793.

MacIntyre, Ferren. "The Top Millimeter of the Ocean." *Scientific American* 230 (May 1974): 62–77.

Ramage, Colin S. "El Nino." *Scientific American* 254 (June 1986): 77–83.

Rasmusson, Eugene M. and J. Michael Hall. "El Nino: The Great Equatorial Pacific Ocean Warming Event of 1982–83." *Weatherwise* (August 1983): 167–175.

THE WEATHER MACHINE

Brownlee, Shannon. "Forecasting: How Exact Is It?" *Discover* (April 1985): 10–16.

Clary, Mike. "Thunderstorms." *Weatherwise* (June 1985): 130–151.

Kerr, Richard A. "Slow Atmospheric Oscillations Confirmed." *Science* 225 (September 7, 1984): 1010–1011.

Molinari, Robert L., et al. "Subtropical Atlantic Climate Studies: Introduction." *Science* 227 (January 18, 1985): 292–294.

Sanders, Ti. *The Weather Is Front Page News.* Icarus Press, 1983.

Snow, John T. "The Tornado." *Scientific American* 250 (April 1984): 86–96.

Webster, Peter J. "Monsoons." *Scientific American* 245 (August 1981): 109–118.

THE WATER CYCLE

Ambroggi, Robert P. "Water." *Scientific American* 243 (September 1980): 101–115.

Brown, Lester R. "World Population Growth, Soil Erosion, and Food Security." *Science* 214 November 27, 1981): 995–1001.

"Facing Geologic and Hydrologic Hazards." *U.S. Geological Survey Professional Paper 1240*-B,

Washington, D.C.: Government Printing Office, 1981.

Gibbons, Boyd. "Do We Treat Our Soil Like Dirt?" *National Geographic* 166 (September 1984): 353–388.

Huppert, Herbert E. "Icebergs: technology for the future." *Nature* 285 (May 8, 1980): 67–68.

Larson, W.E., F.J. Pierce, and R.H. Dowdy. "The Threat of Soil Erosion to Long-Term Crop Production." *Science* 219 (February 4, 1983): 458–464.

Revelle, Roger. "Food and Population." *Scientific American* 231 (September 1974): 161–170.

MAKING WAVES

Begley, Sharon, John Carey, and Elisabeth Bailey. "The Vanishing Coasts." *Newsweek* (September 24, 1984): 74–77.

Bower, Bruce. "Ancient Quake Shakes Up the Past." *Science News* 126 (August 18, 1984): 100.

Fox, William T. *At the Sea's Edge; An Introduction to Coastal Oceanography for the Amateur Naturalist*. Englewood Cliffs, N.J.: Prentice Hall, 1983.

Keller, Jorg. "Did the Santorini eruption destroy the Minoan world?" *Nature* 287 (October 30, 1980): 779.

Kerr, Richard A. "Wither the Shoreline?" *Science* 214 (October 23, 1981): 428.

Lynch, David K. "Tidal Bores." *Scientific American* 247 (October 1982): 146–156.

Myles, Douglas. *The Great Waves*. New York: McGraw-Hill Book Co., 1985.

TREASURES FROM THE SEABED

Anderson, Roger N. *Marine Geology; A Planet Earth Perspective*. New York: John Wiley & Sons, Inc., 1986.

Bonatti, Enrico. "The Origing of Metal Deposits in the Oceanic Lithosphere." *Scientific American* 238 (February 1978): 54–61.

Borgese, Elisabeth Mann. "The Law of the Sea." *Scientific American* 248 (March 1983): 42–49.

Edmond, John M., and Karen Von Damm. "Hot Springs on the Ocean Floor." *Scientific American* 248 (April 1983): 78–93.

Gass, Ian G. "Ophiolites." *Scientific American* 247 (August 1982): 122–131.

Norman, Colin. "Interior Slashes Offshore Oil Estimates." *Science* 228 (May 24, 1985): 974.

Rona, Peter A. "Mineral Deposits from Sea-Floor Hot Springs." *Scientific American* 254 (January 1986): 84–92.

ENERGY FROM THE OCEAN

Beer, Tom. *Environmental Oceanography; An Introduction to the Behavior of Coastal Waters*. Elmsford, N.Y.: Pergamon Press, Inc., 1983.

Craxton, R. Stephen, Robert L. McCoy, and John M. Soures, "Progress in Laser Fusion." *Scientific American* 255 (August 1986): 68–79.

Fisher, Arthur. "Fusion milestone." *Popular Science* 228 (April 1986): 8–12.

_____. "Norsk wave power." *Popular Science* 228 (March 1986): 16.

_____. "Tidal generator." *Popular Science* 227 (August 1985): 12–14.

Moretti, Peter M., and Louis V. Divone. "Modern Windmills." *Scientific American* 254 (June 1986): 110–118.

Penney, Terry R. and Desikan Bharathan. "Power from the Sea." *Scientific American* 256 (January 1987): 86–92.

Peterson, Ivars. "Energy for life among the waves." *Science News* 131 (March 21, 1987): 183.

Scott, David. "Arrowhead windmill." *Popular Science* 228 (April 1986): 98–99.

Smith, Norman. "The Origins of the Water Turbine." *Scientific American* 242 (January 1980): 138–148.

HARVESTING THE SEA

Beddington, John R. and Robert M. May. "The Harvesting of Interacting Species in a Natural Ecosystem." *Scientific American* 247 (November 1982): 62–69.

Brown, O.B., et al. "Phytoplankton Blooming Off the U.S. East Coast: A Satellite Description." *Science* 229 (July 12, 1985): 163–167.

Conover, R.J., A.W. Herman, S.J. Prinsenberg, and L.R. Harris. "Distribution of and Feeding by the

Copepod *Pseudocalanus* Under Fast Ice During the Arctic Spring.'' *Science* 232 (June 6, 1986): 1245–1247.

Cromie, William J. *Exploring the Secrets of the Sea.* Englewood Cliffs, N.J.: Prentice Hall, 1962.

Eastman, Joseph, T., and Arthur L. DeVries. ''Antarctic Fishes.'' *Scientific American* 255 (November 1986): 106–114.

Fisher, Arthur. ''Deepest ocean plant life.'' *Popular Science* 226 (April 1985): 16.

Goreau, Thomas F., Nora I. Goreau, and Thomas J. Goreau. ''Corals and Coral Reefs.'' *Scientific American* 241 (August 1979): 124–136.

Isaacs, John D.,and Richard A. Schwartzlose. ''Active Animals of the Deep-Sea Floor.'' *Scientific American* 233 (October 1975): 85–91.

Shlesinger, Y., and Y. Loya. ''Coral Community Reproductive Patterns: Red Sea Versus the Great Barrier Reef.'' *Science* 228 (June 14, 1985): 1333–1335.

Valentine, James W., and Eldridge M. Moores. ''Plate Tectonics and the History of life in the Oceans.'' *Scientific American* 230 (April 1974): 80–89.

Woche, Wirtschafts. ''Sea Riches: What Future? *World Press Review* (November 1984): 23–25.

A SEA CHANGE

Carrigan, Charles R., and David Gubbins. ''The Source of the Earth's Magnetic Field.'' *Scientific American* 240 (February 1979): 118–130.

Hansen, J.E., ''Global sea level trends.'' *Nature* 313 (January 31, 1985): 349–350.

Hsu, Kenneth J. ''When the Black Sea Was Drained.'' *Scientific American* 238 (May 1978): 53–63.

Meyer, Alfred. ''Between Venice & The Deep Blue Sea.'' *Science 86* 7 (July/August 1986): 51–57.

Pimentel report. ''Ozone in the stratosphere.'' *Environmental Science & Technology* 20 (November 4, 1986): 328–329.

Raloff, Janet. ''Ozone Depletion's New Environmental Threat.'' *Science News* 130 (December 6, 1986): 362–363.

Revelle, Roger. ''Carbon Dioxide and World Climate.'' *Scientific American* 247 (August 1982):

35–43.

Stanley, Stephen M. ''Mass Extinctions in the Ocean.'' *Scientific American* 250 (June 1984): 64–72.

Steinhorn, Ilana, and Joel R. Gat. ''The Dead Sea.'' *Scientific American* 249 (October 1983): 102–109.

Tuck, A.F. ''Depletion of Antarctic ozone.'' *Nature* 321 (June 19, 1986): 729–730.

Turco, Richard P., Owen B. Toon, Thomas P. Ackerman, James B. Pollack, and Carl Sagan. ''The Climatic Effects of Nuclear War.'' *Scientific American* 251 (August 1984): 33–43.

Weisburd, Stefi. ''One ozone hole returns, another is found.'' *Science News* 230 (October 4, 1986): 221.

_____. ''Pole's ozone hole: Who NOZE?'' *Science News* 130 (October 25, 1986): 261.

Woodwell, George M. ''The Carbon Dioxide Question.'' *Scientific American* 238 (January 1978): 34–43.

THE POLLUTION SOLUTION

Bascom, Willard. ''The Disposal of Waste in the Ocean.'' *Scientific American* 231 (August 1974): 16–25.

Begley, Sharon, and Mary Hager. ''Acid Rain's 'Fingerprints'.'' *Newsweek* (August 11, 1986): 53.

Cousteau, Jacques. '''We Face a Catastrophe' If the Oceans Are Not Cleaned Up,'' *U.S. News & World Report* 98 (June 24, 1985).

Dickson, David. ''Europe Struggles to Control Pollution.'' *Science* 234 (December 12, 1986): 1315–1316.

Hollister, Charles D., Richard Anderson, and G. Ross Heath. ''Subseabed Disposal of Nuclear Wastes.'' *Science* 213 (Spetember 18, 1981): 1321–1325.

Kao, Timothey W., and Joseph M. Bishop. ''Coastal Ocean Toxic Waste Pollution: Where Are We and Where Do We Go'' *USA Today* 114 (July 1985): 20–23.

Likens, Gene E., Richard F. Wright, James N. Galloway, and Thomas J. Butler. ''Acid Rain.'' *Scientific American* 241 (October 1979): 43–51.

Maranto, Gina. ''The Creeping Poision Underground." *Discover* (February 1985): 75–78.

Mosey, Chris. "Fish life suffering in polluted Kattegat." *Arkansas Democrat* (November 9, 1986): 27A.

Mullen, Peter, W. "Toxicology of acid rain." *Environmental Science & Technology* 20 (November 3, 1986): 211.

Revkin, Andrew C. "Is There a Balance of Nature?" *Science Digest* 41, 98 (October 1985):.

Sun, Marjorie. "Ground Water Ills: Many Diagnoses, Few Remedies." *Science* 232 (June 20, 1986): 1490–1493.

_____. "OTA Enters Inflamed Debate on Ocean Incineration." *Science* 233 (August 29, 1986): 934.

Tarbuck, Edward J., and Frederick K. Lutgens. *Earth Science*. Bellevue, Wash.: Merril Press, 1982.

Watson, Russel, Debbie Seward, Ruth Marshall, Scott Sullivan, and Frisco Endt. "The Blotch on the Rhine." *Newsweek* (November 24, 1986): 58–60.

Index